THE LAST FOREST

THE LAST FOREST

THE AMAZON IN THE AGE OF GLOBALIZATION

MARK LONDON AND BRIAN KELLY

RANDOM HOUSE | NEW YORK

Published in the United States by Random House,
an imprint of The Random House Publishing Group,
a division of Random House, Inc., New York.

RANDOM HOUSE and colophon are registered
trademarks of Random House, Inc.

ISBN 978-0-679-64305-0

LIBRARY OF CONGRESS CATALOGING-IN-PUBLICATION DATA
London, Mark.
The last forest: the Amazon in the age of globalization /
Mark London and Brian Kelly.
p. cm.
Includes bibliographical references.
ISBN-10: 0-679-64305-2
ISBN-13: 978-0-679-64305-0
1. Rain forests—Amazon River Region—Management.
2. Forest management—Amazon River Region. 3. Rain for-
est ecology—Amazon River Region. I. Kelly, Brian
II. Title.
SD418.3.A53L66 2007   333.750981′1—dc22
2006046466

Printed in the United States of America on acid-free paper

www.atrandom.com

24689753

Book design by Barbara M. Bachman

FOR

The London Kids (Jana, David, and Scott)

AND

The Kelly Kids (Jack, Laura, and Danny)

PREFACE

In November 1980, to prepare for the first of two one-hundred-day expeditions to the Amazon that would provide the content for our first book, *Amazon*, we wended among the skyscrapers in the frenetic business district of São Paulo, Brazil, to interview a man we hoped would prove worthy of his title: President of the Association of the Impresarios of the Amazon. The burgeoning international environmental movement had marked this man as a candidate for public enemy number one. João Carlos Meirelles was the leader of a coterie of entrepreneurs based in the developed southern part of Brazil who were supposedly responsible for the accelerating deforestation of the Amazon. Cattle ranching, often on spreads nearly as large as some U.S. states, was their industry of choice.

With thick brows and a mustache that he twisted with a touch of wax at the ends, a vested pin-striped suit, a watch chain at his waist and a silver-headed walking stick in hand, he was the perfect embodiment of what we imagined a Latin American aristocrat should look like. With a stern, intense demeanor, he told us, "I have a passion, not a job."

As hospitable as he was that day, he surprised us with his brusqueness in conversation. He accused us of trespassing in areas of concern where we didn't belong. "It is the Europeans who are so concerned with our jungle, the European magazines and newspapers that write about trees being cut and stir up ecologists," he said. "They have time to worry about it and they cause a problem for us. There was no ecology movement when you were developing your country, but now everyone worries about us."

Our ignorance of the Amazon, he assured us, dwarfed our knowledge, a condition he found prevalent in Europeans and North Americans. He claimed the impresarios who had alighted in the Amazon shared the same ambition and capability as the moguls who'd developed America's frontier. He characterized the impresarios as national heroes.

We told Meirelles we had heard that one third of the Amazon rain forest had been deforested, and pressed him to identify "heroism" in that statistic. Meirelles scoffed. "That jungle is so big, I don't think one percent is gone. But more will come down; it has to. We do not have the time to stop and study it. I'm telling you, it's impossible to stop the occupation of the Amazon."

We left São Paulo to find out what was happening in the Amazon. At the time, information about the region was scarce, unreliable, and dated. The Amazon was still a faraway place shrouded in the myth of a remote, uninhabitable world. But the reality evidently was very different, as frenzied occupation of the Amazon was ongoing. Who was there? Why were they there? What were they doing? And what were the consequences?

We sought to answer these questions in a book about a mélange of people living in the Amazon. An Indian tribe, teetering on extinction, invited us to dinner on our Thanksgiving holiday. We walked among armed squatters, who were desperate to protect their land from gunmen hired to kill them. We visited the mythic gold mine of Serra Pelada and fabled projects of biblical proportions—Henry Ford's gargantuan rubber plantation and D. K. Ludwig's industrialized tree farm the size of Massachusetts and Rhode Island combined. We rode buses with immigrants from the settled areas of Brazil, who were following rumors of available land they could call their own, where they could start a new life. A frontier had opened, where lost dreams littered some landscapes and hope abounded on others, and all the while the rain forest suffered fatal wounds. In the end, we agreed with Meirelles that human occupation of the Amazon on a large scale was inevitable—and its scars were indelible. It seemed that saving the Amazon forest while addressing the needs of the people living there would require a miracle.

In the spring of 2003, before returning again to the Amazon, we first visited an ornate colonial government building in São Paulo to see João Carlos Meirelles, then sixty-eight years old, serving in the cabinet of the state government. He hadn't lost any of his passion, although when we asked him to reflect back on our conversation twenty-three years earlier, the serious mien we remembered softened. Did history vindicate or repudiate his self-assuredness of 1980? He laughed. "We made many mistakes; you know that. And we have to accept this with humility by learning from them."

The mistakes Meirelles referenced had fueled the international environmental movement, which adopted the deforestation of the Amazon as a cause célèbre in the 1980s and 1990s. A land area larger than France had been cleared of trees. Scores of indigenous tribes had lost out in the struggle to adapt to the modern era. Scientists lamented the loss of the biodiversity (a term that had been coined since our last visit) of the richest rain forest before it had even been catalogued. Brazil itself had been attacked by world leaders for failing to preserve a global resource.

Meirelles lamented that his organization had contributed to the problem, as the impresarios "didn't build anything sustainable, to last." He also pointed out that many of the hopeful immigrants we had met had been lured to the Amazon by the prospect of employment on public projects such as road building, but these workers had been "abandoned" once the jobs were done. The Amazon was an exotic place to Brazilians from other parts of the country, and many found it difficult to adapt to the inhospitable climate and seemingly non-arable soil.

Despite this litany of failure, Meirelles was as enthusiastic as we had remembered him. He lauded the changeover from the military government in the mid-eighties to an entrenched democratic system, pointing out that the newly elected populist president Luiz Inácio Lula da Silva ("Lula") had been in jail for union organizing activities during our first visits. Meirelles's optimism about the Amazon arose not only from the lessons learned but from the new tools available. "We are in a new phase of technology," he said. "We have satellite information that allows us to determine where agriculture might be successful and where it shouldn't be attempted; we have information about the soil that allows us to grow in areas we never thought were arable; we can protect the water and soil from erosion; we can breed cows more efficiently and have pastures with high yields, which means less land will have to be cleared."

He added, "We have a greater appreciation of the environment. The challenge now is to define what we have to protect, and to protect it indefinitely."

The Amazon has changed considerably in the more than twenty-five years since the visits for our first book, and *The Last Forest* traces this transformation. Over this time span, the world's concern for the fate of the rain forest and the world's recognition of its importance to the global environment have substantially increased. Yet the conventional wisdom of what it will take to save this resource is stuck in time.

To say that the Amazon is being destroyed misstates what is happening. The Amazon is being changed. At times, deforestation results in the creation of arable land, capable of sustaining a family, or a group of families that becomes a community. At other times, deforestation is the first step toward building the breadbasket of tomorrow, providing millions of people with new sources of protein and Brazil with foreign currency to support desperately needed social programs. And at other times, deforestation is the first step in a process that yields jobs, which provide capital, which provides a sense of security and a commitment to the future. But there are many times when deforestation provides nothing more than needless destruction of one of the world's greatest natural resources and the indigenous people who inhabit it.

Saving the Amazon now requires saving the people who live in the Amazon—more than twenty million of them—by making educated and collaborative judgments about development and preservation. Preaching to Brazil from London, Paris, or Washington about what its citizens can and cannot do or can and cannot touch causes nothing but resentment. To save the Amazon, we must recognize this region not as an exotic wilderness but as one of the few frontiers left on earth where, as Meirelles said, opportunities abound. The Amazon presents a challenge in nation building for Brazil, an opportunity to enfranchise its citizenry and create a positive sense of national participation, an opportunity rarely seen anymore in the developing world. The challenge also presents Brazil with an opportunity to discharge an enormous responsibility among the family of nations—to serve as guardian of the world's largest repository of biodiversity and source of fresh water, as well as a crucial stabilizer of its climate. If it succeeds it will deserve international gratitude.

Why does any of this matter? The easy answer is, because of the substantiated predictions about changes to the world's climate and predictions about the rapid depletion of the world's natural wealth if the rate of deforestation continues. But the more complex answer also hints at a solution. In his post-9/11 "wake-up call" speech, Blairo Maggi, the governor of the Amazon state of Mato Grosso, warned that as the developed world reacted to the violence of the day it would lose sight of the cause of that violence. Terrorism is a symptom of alienation, he argued. Degradation of the environment is its cousin. People degrade the environment because they do not perceive themselves as invested in its future, not because they hate trees or clean water. Reasonable people do not deforest their backyards, nor destroy what they own. The environmental problems in the Amazon have arisen because the political and economic system in Brazil, as well as globally, has failed to create a sense of interest in the clear connection between an individual's well-being and the well-being of the environment. Or because commercial self-interests often trump the common interest in protecting natural resources. That must change.

The Amazon today is a world apart from its historic stereotype. There's a struggle going on among its inhabitants to turn opportunity into a productive world and to create a sense of belonging to that world, a sense of entitlement to a place in it. Resistance to these efforts must be broken. The outcome of this struggle will influence the fate of the world's environment and millions of people seeking a chance to participate in the global economy.

It takes time, sometimes more than twenty-five years, to identify problems and perceive solutions by recognizing how competing interests need to accommodate one another. In this book, we attempt to show what it will take to save the Amazon, and maybe even a larger universe. It is not too late, although not by much.

CONTENTS

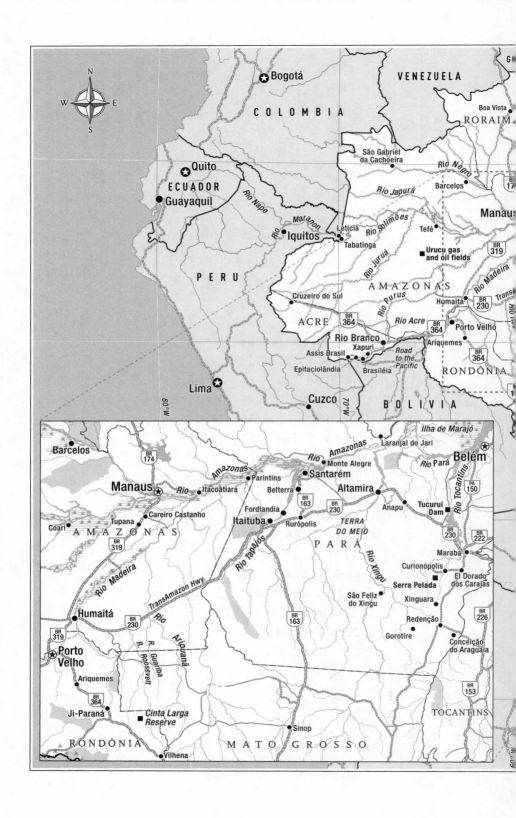

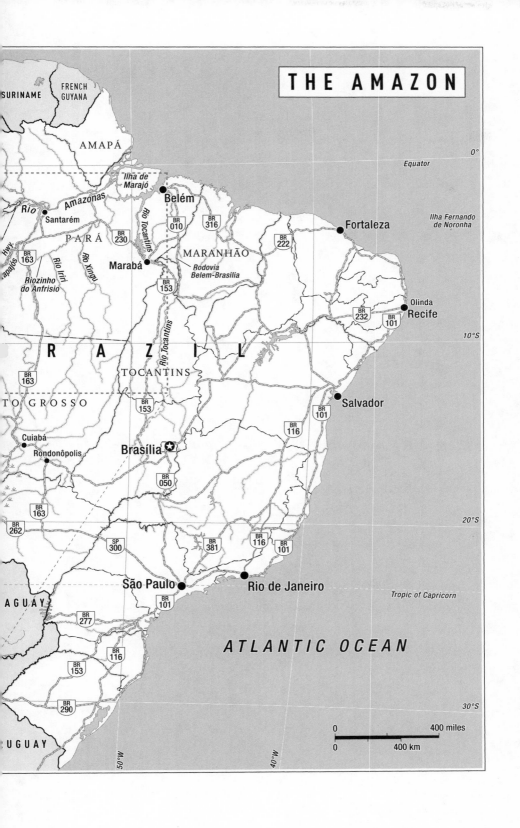

THE LAST FOREST

CHAPTER 1

AN UNEXPECTED BEGINNING

—

From the time he was a small boy, Nelsi Sadeck heard about the cave paintings out beyond the sandstone ridges that fingered their way down toward the north bank of the Amazon River. He and his friends already were well acquainted with the paintings high on the cliffs closer to town, really nothing more than lines of dull colors against gray rock. Because they knew only their small river town of Monte Alegre and its environs, they never knew just how special these paintings were in the region. Along the main Amazon for two thousand miles upstream, there was nothing taller than a tree, except for these rocks. And other than the chocolate brown water of the swift-moving river and the sentries of green trees guarding the riverbank, there was no color. Anyone from anywhere else in the region would have known Sadeck had seen something special up in those rocks, but for him, these were nothing more than local artwork.

Then, in the early 1970s, Sadeck started to hear stories about other paintings, ones scattered randomly in caves hidden in the small hills farther inland. He had heard by then from enough visitors that there was nothing else like them in the vast green sea of Amazon forest. So, he went exploring. The paintings he found were simple red and yellow renderings of animals and people, childish and exuberant—primitive depictions of spiders, frogs, owls, and giant snakes. He saw stick figures of men and women and a rail-thin cow with horns. Bright suns and handprints mixed with geometric spirals and concentric squares. One painting looked to be a calendar, a six-by-eight-foot rectangle lined by precise squares, some with X's through them. No record exists of the society that created these paintings, so it's just as easy to speculate that they depict a game of chess or tic-tac-toe or a calendar. There are seven such cave sites, Sadeck came to learn, with hundreds of paintings. Still, thirty-five years after he first saw the caves, Sadeck believes others may exist.

Sadeck didn't discover these caves. More than a century before Nelsi Sadeck went exploring, Alfred Russel Wallace, a naturalist of Darwinlike importance, came to the Amazon to survey the flora. He was a botanist, not an anthropologist. And though Wallace's scientific discoveries provided much of the foundation for ecological scholarship of the Amazon, he was more concerned with nature and missed the main event—the historic record of humans in the area. Not for another 140 years would the significance of these painting be appreciated.

In 1988, a scholar named Anna Roosevelt came to Monte Alegre and asked to see the caves. By this time, Sadeck, a high school ecology teacher, had become the caves' caretaker and guide, the man to see in Monte Alegre if you were the odd off-the-beaten-path tourist or enterprising academic. Monte Alegre lies about sixty miles downriver from the regional hub of Santarém, which is the largest city between Belém, at the mouth of the Amazon River where it meets the Atlantic, and Manaus, about four hundred miles upstream. Santarém sits on the south bank of the Amazon at the river's intersection with the crystal blue Tapajos River, whose waters run side by side with the muddy Amazon for about ten miles before the brown swallows the blue. The journey from Santarém to Monte Alegre is by car to the river's edge, then by barge, and then by a waiting flatbed truck that Sadeck hires for the visit.

Sadeck accompanied Dr. Roosevelt the first time she saw the caves, and he has guided many of her subsequent trips. His enthusiasm hasn't diminished over time. "This is a magical place, and Anna has explained that to the rest of the world," he told us. "This gives history to the Amazon."

These caves actually provide a lot more than history. Roosevelt's findings have revolutionized our understanding of the Amazon's place in history, in Brazil, and in the rest of the world. Her theories that humans once thrived in the Amazon (hinted at by anthropologists who had preceded her but without the startling evidence of these caves) have radically altered political and scientific perceptions of this rain forest—perceptions not only of its past but of its future.

In forging this academic breakthrough, Anna Roosevelt is a worthy heiress of her great-grandfather President Theodore Roosevelt, whose courageous exploration of the heart of the forbidding Amazon rain forest is well documented by Candice Millard in *The River of Doubt*. TR traveled the length of a river that now bears his name (the Rio Roosevelt), braving the stew of tropical horrors: diseases, insects, isolation, and hostile indigenous tribes. In the end, the jungle won out, as TR, the epitome of the robust American, died several years after the Amazon had destroyed his health.

Anna Roosevelt, a former curator of the Field Museum in Chicago and now an anthropology professor in the University of Illinois system, made her first discoveries about the Amazon in the early 1980s in Cambridge, Massachusetts—at Harvard's Peabody Museum. Following a lead she had come upon in a 1960 article about Amazon archaeology, she traveled to Harvard to examine the long-neglected collection of pottery and shells donated by Charles Frederick Hartt, a promising geologist who made several trips to the Amazon in the 1860s. As Hartt didn't have sophisticated dating equipment available to him, he, like Wallace, could not appreciate the significance of what he had come upon.

Roosevelt, using radiocarbon technology, concluded that Hartt's samples were over six thousand years old—"at the time the earliest

date for pottery in the New World." That the oldest trace of ceramic society on the continent was found in the Amazon, she wrote, set "the
stage for the revision of Amazon culture history, a process that was to
reverberate in New World culture history as a whole."

The age of the pottery showed a ceramic society in the Amazon at
least three thousand years before the Amazon was thought to have
been settled and, more important, it was older than any Andean pottery that had been found. That meant the Amazon wasn't settled by
"ceramic-age agricultural people from the Andes," the prevailing theory throughout the twentieth century. Amazonians came first, or at
least they developed independently of Andean society.

Roosevelt wasn't done in the Peabody. She also surmised that a
"pile of yellow pages" must have been "Hartt's long-lost book," a man-
uscript of his findings that one of his students had sent to Harvard
after Hartt's death in Rio from yellow fever in 1878. Hartt described
finding spear points at a site near Monte Alegre. Roosevelt was intrigued because she didn't recall such relics at any ceramic-age sites
she had explored or read about. She sensed the presence of an even
older society.

Both Wallace and Hartt had written about the cave paintings, and
Roosevelt appreciated the uniqueness of such a place in Amazonia.
Charles C. Mann describes the site in his book *1491*: "Wide and shallow and well lighted, Painted Rock Cave is less thronged with bats than
some of the other caves. The arched entrance is twenty feet high and
electric with gaudy petroglyphs. Out front is a sunny natural patio,
suitable for picnicking, that is edged by a few big rocks. During my
visit I ate a sandwich atop a particularly inviting stone and looked
through a stand of peach palms to the water seven miles away and the
forest between. The people who created the petroglyphs, I thought,
must have done about the same thing." Roosevelt focused on the thick
black dirt on the pathway just outside the entrance to *Caverna da Pedra
Pintada* (Painted Rock Cave), a stunning site, albeit infested by wasps
on the day of our visit. She hypothesized that by excavating the soil
around the petroglyphs she would learn more about the people who

had drawn them. In 1991 and 1992, her team unearthed "30,000 stone artifacts, pigment, and many thousands of burn nuts, seeds, shells and bones." The dating process showed proof of human habitation at this site from between 11,200 and 10,000 years ago.

Until Roosevelt published her findings in a series of articles for *Science* magazine, scientists had adopted the theory that South American civilization had drifted down from the north. The skeletons found in Clovis, New Mexico (hence, Clovis Man), were thought to be the forefathers of the original tribes of South America. It was thought that when they ran out of food, they headed south and eventually east from the Andes to Amazonia. But Roosevelt's findings proved that there was a contemporary culture in the Amazon before Clovis Man traveled southward. This discovery of a contemporary parallel universe in the south, as well as the discovery that the continent's oldest pottery samples are in the Amazon, roiled the world of science in the 1990s. The debate hasn't subsided.

The controversy boils mostly because if Roosevelt's theories are right, the long-held theories of how the Amazon was settled must be wrong. The person most identified with the opposing viewpoint is Betty Meggers of the Smithsonian Institution, a woman of legendary status in Amazon scholarship. (When we first wrote about the Amazon, we were told that our work would have no credibility without input from Meggers; we placed a visit to her musty artifact-cluttered office at the Smithsonian at the top of our list.)

In 1948, she and her archaeologist husband, Clifford J. Evans, began fieldwork on Marajó Island, an island the size of Switzerland that clogs up the mouth of the Amazon near Belém like a fist in a trickle of water. Meggers and Evans concluded that the scarcity of evidence of a settled culture on Marajó meant the early Amazonians were a nomadic culture. The Amazonian environment, they argued persuasively, isn't conducive to anything but slash and burn agriculture: burn the trees, plant in the nutrient-rich debris until it washes away, and then move on. The two academics published their seminal findings in *American Anthropologist* in 1954: "Even modern efforts to implant civilization in the South American tropical forest have met with defeat, or survived only with constant assistance from the outside. In short, the environmental potential of the tropical forest is sufficient to allow the evolution of culture to proceed only to the level represented by [slash and burn farmers]; further indigenous evolution is impossible, and any more highly evolved culture attempting to settle and maintain itself in the tropical forest environment will inevitably decline to the [slash and burn] level."

This last sentence defined twentieth-century scientific thought about human occupation of the Amazon, until Sadeck and Roosevelt visited Painted Rock Cave one afternoon in 1988. Meggers postulated that the Amazon had to remain untouched or it would be destroyed. That was history's lesson. There was no middle ground, no precedent for any sustainable development.

Meggers was writing on a tabula rasa. No scholarship of any note had been published about the early settlement of the Amazon; no thought had been given to human history in the Amazon. And she was theorizing the obvious: the physical climate is so recognizably hostile that Meggers had a readily receptive audience for her explanation as to why there were no long-term settlements and only sparse nomadic populations. The environment stymied civilization. Paul Richards, an American biologist, provided scientific support, writing in the early 1950s that the soil in the Amazon was infertile and contained a shallow, albeit dense, level of nutrients that washed away in the heavy rains after several years. Without the prospect of permanent agriculture and the ability to create surpluses, there was no possibility of creating stable and lasting institutions or meaningful social, political, or economic systems.

Meggers expanded on her "touch it and you'll destroy it, just as everyone else has" theory in her 1971 book, *Amazonia: Man and Culture in a Counterfeit Paradise*. She stated that Amazonia "with all its wonderful intricacy [is] like a castle in the sand." A more influential book has never been written on the Amazon. It was the right message at the right time, just when the global environmental movement was in formation and in search of a great cause. Worldwide interest in the Amazon was beginning to stir and Meggers's message fit the growing global concern for environmental preservation. The messenger was also particularly attractive: not only a courageous woman in a man's field but also an American who embraced the work of local anthropologists. Her stature and subsequent academic support for her conclusions, coupled with a prevailing antidevelopment attitude in the scientific community, made the Amazon ethically "untouchable."

In 1975, another important book added a social planning dimension to the ecological argument. Two scientists from the World Bank, Robert Goodland and Howard Irwin, published *Amazon Jungle: Green Hell to Red Desert?* using Meggers's theory to criticize development projects in the Amazon. "The purpose of this study," they wrote, "is to show what little is known of this immense but fragile area, to relate what is being done, to predict what the environmental results may be, and to suggest some means of averting or at least blunting predictably vast and tragic consequences that loom ahead." When those "vast and tragic consequences" began to appear in the Amazon in the late 1970s—for example, mammoth forest clearings for unproductive cattle ranches or the environmental destruction wrought by the TransAmazon Highway and scores of failed settlements along its route—evidence mounted in favor of leaving the Amazon alone. No productive use could be made of it, it seemed.

Yet as pioneers began to move into the region from other parts of Brazil and experimented with seed varieties and crop rotations, and as scientists, aided by improved tools, did fieldwork in the region, the base of knowledge about the soil's capability and past experience began to build. Slowly, the evidence began to point toward other postulates. One of the early skeptics of Meggers's theory was Donald Lathrap, an anthropologist from the University of Illinois. His work in the Peruvian Amazon in the late 1960s led him to theorize that the Amazon had actually been populated by large settled groups capable of sustaining agriculture. He further theorized that civilization had moved from the Amazon to the Andes, not vice versa. There was agronomic support for his theories from Robert Carneiro, who challenged the view that the environment inhibited soil fertility, and Carl Sauer, who concluded that indigenous people had "extensively modified the tropical forest" for sustainable agriculture. Nevertheless, Lathrap encountered a skeptical audience comfortable with the view that the Amazon never had been and never should be settled. He died in 1990 before his ideas gained currency.

However, Lathrap broke through a barrier, and others, armed with empirical research, followed. William Denevan, a geographer at the University of Wisconsin, discovered a site in lowland Bolivia that provided the basis for his 1992 paper, "The Pristine Myth: The Landscape of the Americas in 1492," in which he argued that vast areas of the modern Amazon rain forest had actually been settlements rather than stopovers for transients in slash and burn societies. That humans could survive this hostile environment, especially with only rudimentary technology, challenged the way the world viewed the Amazon. Perhaps, scientists began to speculate, the Amazon was meant to be inhabited all along. The beginnings of acceptance of long-standing human occupation of the Amazon started to seep into the public debate.

While Denevan was analyzing his findings, Roosevelt was beginning to publish the results from her visits to Marajó Island, a site she had visited before Painted Rock Cave. As the research tools available in the 1980s were superior to those available to Meggers and Evans in the 1940s, Roosevelt was able to excavate more evidence to support her conclusions. And those conclusions essentially blasphemed Meggers's scholarship: Marajó, according to Roosevelt, was home to an advanced civilization of up to one hundred thousand people that lasted over a thousand years. Meggers, she argued, had had it all wrong. People had flourished in this place.

Meggers did not react well. She accused Roosevelt of incompetence. She claimed that Roosevelt had unearthed not the site of a large population but scattered sites of small groups that appeared over a long period of time. Roosevelt, in turn, called Meggers a tool of the CIA, an accusation that grabbed headlines but added little substance to the debate. Roosevelt pressed on, continuing upriver to Monte Alegre, where she made additional findings at Painted Rock Cave that solidified her standing.

Personalities aside, Meggers came under siege from many quarters. Technology played a large role. Soil analysis has shown that perhaps as much as 10 percent of the Amazon consists of *terra preta* (black earth), with high levels of organic matter and carbon—persuasive evidence of human occupation. The density of the jungle, which had long been considered as proof of the jungle's hostility to permanent settlements, was in many areas hiding evidence to the contrary. Anthropologists couldn't see remnants of past civilizations, so they theorized that the civilizations had never been there. Susanna Hecht, a researcher in many Amazon disciplines, analogized this misperception to John Muir's gazing "upon the majesty of Yosemite." She wrote, "What he took as a wilderness was to other eyes an agricultural landscape formed of trees and tubers which his own conception of agriculture, and his own conception of nature, could not comprehend. This area, so majestic in its beauty and its vegetation, had been both human artifact and habitat." So was the Amazon.

We were once told by Robert Goodland, as authoritative a source as there was in the 1970s, "You can think of it as a desert covered with trees." That claim was considered indisputable when we wrote our previous book. Now it is wrong, perhaps on a grand scale. Today the debate is not whether large areas of soil are fertile (indeed, they are) but whether settlers in the area deliberately and repeatedly improved the soil to this degree, or whether the soil's improvement was a by-product of repeated settlements.

Large areas of the Amazon have been populated over time, and, more startling, large areas of the Amazon have been modified by man. The long-held belief that the Amazon is a lost Eden has been shattered. Journalists' writings of their travels through the region twenty years ago, such as those by Andrew Revkin (*The Burning Season*) and Jonathan Kandell (*Passage Through El Dorado*), mention the *terra preta* findings and Denevan's views but reveal a skittishness in giving these views too much credence. These were anti-environmentalist theories at a time when this movement still occupied the cocoon of moral absolutism.

Scientists now are examining this transformation of the *terra preta* soil from barren to fertile, trying to learn if the process was intentional or not and how to replicate it, since the indigenous groups apparently achieved an enviable level of environmental harmony. As Mann has written, "The new picture doesn't automatically legitimize paving the forest. Instead, it suggests that for a long time big chunks of Amazonia were used nondestructively by clever people who knew tricks we have yet to learn."

Civilization apparently thrived throughout this expansive and seemingly uninhabitable place—it is more than two thousand miles from Roosevelt's work on Marajó to Denevan's in Bolivia. In 2003, Michael Heckenberger of the University of Florida led a team to the headwaters of the Xingu River and found evidence of the most sophisticated society yet discovered. About eight hundred years ago, his team concluded, the area was widely inhabited. There was a planned network of nineteen settlements—each with twenty-five hundred to five thousand people—built around large circular plazas with roads leading from them. The settlements had roads with curbs, defensive moats, canals, and bridges. The towns also had managed forests and cultivated fields.

These findings still cause polarizing debate. Yet, more and more research continues to emerge that supports the theory that people lived in the Amazon for much longer than originally thought, at a higher standard than originally thought, and in much more adaptive ways than originally thought.

Meggers's theory appears increasingly fragile. Now into her eighties, she has not relinquished her claim that the Amazon is a counterfeit paradise, inhospitable to civilization. She wrote recently, "Adherence to the lingering myth of Amazonian empires not only prevents archaeologists from reconstructing the prehistory of Amazonia, but makes us accomplices in the accelerating pace of environmental degradation."

There is a religiosity to this debate. Meggers's view provides historic justification for what environmentalists want most of all: to have the place left alone. But it was never left alone. For more than ten thousand years, much of the rain forest hasn't been pure, primeval, or primitive. If an earlier civilization successfully settled here before the Europeans came with their diseases and murderous ways, then why can't it happen again? The world's environment as well as the future of Brazil and its 190 million citizens may depend on the answer to this question.

CHAPTER 2

MYTH GIVES WAY TO SCIENCE

—

The discoveries surrounding the permanent settlements, and the growing appreciation for the agronomic technology behind *terra preta,* cast a different light over the history of Amazonia. In today's lexicon, it's universally acknowledged that no one from Spain or Portugal, or even Clovis, New Mexico, "discovered" the Amazon. (One of Columbus's captains navigated the mouth of the river in 1500, and is given credit for the first European sighting, but thousands of people had already "discovered" the region and were living there.)

The knowledge that humans first arrived more than ten thousand years ago has reignited debate over the origin of these early Amazonians. At least three diffusion theories vie to explain how people got there in the first place—some people say the first Amazonians came from Africa, others espouse the reverse Kon-Tiki raft route from Asia, and others stubbornly insist that Clovis Man was the first settler. No matter their origin, tens of thousands of people already lived in the Amazon when a bunch of bedraggled Spaniards, hungry and climate-beaten, "discovered" the river on February 11, 1542. Recent anthropological finds suggest that these people lived with knowledge about their environment and in a healthy symbiosis that we have yet to re-create. Through violence and infectious diseases, the conquistadors obliterated societies that had been productive while sustaining their environment.

History has bestowed worthy heroes in some of the world's more mythic places, such as Hillary atop Everest or Lewis and Clark in the American West. On this scale, the Amazon receives the short end of the stick. There haven't been many more odious characters within the pages of history than Gonzalo Pizarro, who led the expedition result-ing in the first European exploration of the Amazon River. The half brother of the Incan conqueror, Francisco, who appointed himself gov-ernor of Peru in Cuzco, Gonzalo Pizarro followed his sibling to Peru from Spain, evidently looking for a handout. He soon made himself a problem-relative. To get rid of him, Francisco named him governor of Quito, almost one thousand miles to the north of Cuzco. Thrilled with this new authority, Pizarro quickly developed a plan to exploit and ex-pand his domain. He had heard rumors of groves of cinnamon trees leading to a lake lined with gold, a place called El Dorado. His cousin, Francisco de Orellana, who had "discovered" Guayaquil in Ecuador, had been experiencing similar stirrings, and the two decided to team up and seek their glory. Pizarro's expedition included, depending on which account you believe, up to 350 soldiers, with 200 of them on horseback; two thousand llamas; two thousand pigs; two thousand dogs supposedly trained to attack Indians; and four thousand Indians, whom the Spaniards treated like dogs. The explorers took axes, hatch-ets, and ropes.

Pizarro waited for his cousin to join the team but finally lost pa-tience and headed out without him over the mountains. He fared about as well as one might expect from someone who had grown up on the hardscrabble plains of Spain and whose road map had been drawn from confessions he'd beaten out of captives along the way, who, to avoid being tortured, claimed they knew the secret passageway to El Dorado. The Spaniards' horses couldn't climb the snow-covered rocks, Pizarro's men suffered through torrential rains, many of his animals died or ran away, and his slaves dropped by the wayside from exhaustion. Once they crossed the Andes and started to descend into the rain forest, the misery only worsened. Orellana, playing catch-up, pushed his men through a forced march. All of his horses died, forcing him to abandon his gear.

Finally, the two units did meet up. Both were maddened by the results of the trip, and they decided to split up, reconnoiter separately, and then reconvene to share intelligence. Pizarro went ahead and actually found a few cinnamon trees but no gold. Angry, he threw half of the remaining Indians to the dogs and burned the other half alive.

Orellana went by boat down the Napo River, embarking on December 26, 1541, to search for tribes he could pillage for food and supplies. He evidently had no intention of rejoining cousin Gonzalo. The historical record caught a minor break here: Orellana brought along Gaspar de Carvajal, who kept a log of this journey. Unfortunately, much of it was intended to be fiction. Carvajal and Orellana conspired to fabricate an account of an unintentional separation from Pizarro, who they figured would behead them if he ever found out what they were really up to. They had decided to ride the river to its mouth; they had no plan to turn back and replenish Pizarro's team, who by this time had eaten the remaining dogs and horses. Little did Orellana and Carvajal suspect that the journey would take eight months.

The first account of a journey down the Amazon, then, is a long self-serving alibi rather than a journal of discovery. Carvajal dwells on his misery to evoke pity from readers who might care that he abandoned Pizarro, who was slogging back to Quito with a scrum of half-dead men. According to Carvajal's diary, one Indian tribe after another attacked them, and at one point an arrow lodged in his eye, blinding him.

This arrow supposedly was shot by Amazons, and it is Carvajal's description of these female warriors that has brought him five hundred years of ridicule. "The Amazons go about naked but with their privy parts covered, with their bows and arrows in their hands, doing as much fighting as ten Indian men," he wrote. And he added, "We saw the Amazons in front of all the Indian men as women captains, fighting so courageously that the Indians did not dare to turn their backs, and if they did the women clubbed them to death before our very eyes."

For a half millennium no one believed Carvajal's descriptions of continuous habitation along long stretches of riverbank. As the primary motivation for his diary was falsification, and as these descriptions would not match the observations of thousands of subsequent travelers, he had been dismissed as little more than a teller of tall tales. Around Santarém, he claimed to have encountered hundreds of canoes, each filled with twenty to forty Indians, and "squadrons on the riverbank dancing about and waving palm branches." Historians concluded that this was fantasy, and dismissed his report as "foolish boasting." The web Carvajal wove to deceive Pizarro succeeded in misdirecting centuries of scholarship. But now that historians believe that there were thousands of people populating the Amazon in orderly settlements, Carvajal is getting a fresh read for hints of truth weaving through his travails. The population may not have been dancing to celebrate his arrival, but he saw lots of people living there. That much now appears to be confirmed.

Orellana and his team reached the Atlantic Ocean on August 26, 1542. He rushed back to Spain and had himself appointed governor of the Amazon, but died of tropical fever before ever sailing back up the river. Carvajal later became archbishop of Lima and died in 1584. By this time, the order of so-called progress in this part of the world had been established: conquistadors would be followed by missionaries, who would be followed by traders. And the native population was annihilated in the process by murder or disease.

It was one hundred years before another written account of a journey appeared, penned by a Jesuit priest, Cristóbal de Acuña, who published *The New Discovery of the Great River of the Amazons* in 1641. Acuña made some useful observations about animal and plant life and didn't seem bent on killing every Indian he saw, but his account, like Carvajal's, is also suspect because of his peculiar motive: to describe a place so idyllic and valuable that the Spanish crown would want to wrest it away from Portugal, to which it had been allotted by the boundary line drawn in the Treaty of Tordesillas in 1494.

Another one hundred years passed before the arrival of the next diarist, a French scientist, Charles-Marie de La Condamine. By this time, the population in situ at the time of Orellana's journey had been decimated. Although La Condamine's excursion provides an idea of what the area looked like in 1745, any reliable observations of the previous two hundred years, when these settlements were thriving, are missing from the historical record.

La Condamine wrote his *Abridged Narrative of Travels Through the Interior of South America from the Shores of the Pacific Ocean to the Coasts of Brazil and Guijana, Descending the River of the Amazon* as a result of an expedition to South America to test Newton's theory that the earth bulged at the equator and flattened out at the poles. La Condamine's method was to measure the length of degrees of longitude at various places on his journey, but once he arrived at the Amazon, he decided to stay and explore. The French scientist made complimentary observations about the skills of the Indians he visited, though he had little use for the people themselves, as he found them lazy and slovenly. Their knowledge of their environment impressed him, and he was the first to report on the existence of rubber trees and quinine bark, foreshadowing recognition of the value of the rain forest's inventory. La Condamine's substantive reports undoubtedly advanced science, but his greatest contribution may have been his appreciation for the indigenous people's knowledge of their environment. A careful reader of his reports would realize that a natural laboratory of vast potential, virtually unknown, was awaiting discovery.

In the mid-nineteenth century, three visitors with powers of observation commensurate with their subjects came to catalogue the Amazon: Alfred Russel Wallace, Henry Walter Bates, and Richard Spruce. Many credit Wallace, a giant among naturalists, as the scientist whose work laid the foundation for Darwin's *On the Origin of Species*. He and Bates set out together for the Amazon in 1848 with the intention of working on a comprehensive study of the region. Yet their perspectives differed and they decided to split up in 1850. Wallace founded the discipline of zoogeography and was interested in theory, while Bates was a meticulous observer and interested in minutiae. Each found success in his own right, Wallace with a precursor theory to Darwin's, and Bates with a list of 14,712 different species, 8,000 of them new to science. (It was fortunate that Wallace focused on the big picture, as his ship caught fire on the return to England and he lost all of his samples, leaving him only with his evolutionary theory of how the samples all fit together.) When Bates read Wallace's paper and then Darwin's, the two writings provided him with a "unifying framework" to explain the complexity of the rain forest. The concept of natural selection explained what Bates saw in "a palatable species [of butterflies] mimicking an unpalatable one" in order to survive attacks from hostile birds. Bates's highly readable account—*The Naturalist on the River Amazons: A Record of Adventures, Habits of Animals, Sketches of Brazilian and Indian Life, and Aspects of Nature Under the Equator, During Eleven Years of Travel*—endures as a masterpiece: his book advanced the state of knowledge about the Amazon more than any other work.

Spruce's contribution was almost as important, although much less celebrated because he did not publish his findings. Wallace, an uncommonly generous scientist, recognized the importance of his colleague's work and arranged for its publication. It is Spruce who deserves credit for understanding the potential of quinine to treat malaria. He also set a standard for sympathetic treatment of indigenous peoples, best demonstrated by his learning to speak twenty-one different languages.

By the first half of the twentieth century, tales of the Amazon shifted from sober scientific inquiry to breathless adventure stories, some true but most not. Theodore Roosevelt's *Through the Brazilian Wilderness*, published in 1914, is one of the tamer and more scholarly works of this time, although the physical challenge of his trip often pushed him to hyperbole. "South America," he wrote, "makes up for its lack, relatively, to Africa and India, of large man-eating carnivores by the extraordinary ferocity or bloodthirstiness of certain small creatures." His depiction of one of those creatures, the piranha fish, is a model of excessive Amazon writing: "They will rend and devour alive any wounded man or beast, for blood in the water excites them to madness. The razor-edged teeth are wedge-shaped like a shark's and the jaw muscles possess great power. The rabid furious snaps drive the teeth through flesh and bone. The head, with its short muzzle, staring malignant eyes and gaping, cruelly-armed jaws is the embodiment of evil ferocity." Notwithstanding his exuberance, Roosevelt possessed a remarkably keen mind, and made a prophetic observation at the end of his book, where he calls the Amazon "the last true frontier." He added, "Surely, such a rich and fertile land cannot be permitted to remain idle, to lie as a tenantless wilderness, while there are such teeming swarms of human beings in the overcrowded, overpopulated countries of the world."

The notion that the Amazon could be an important human outpost was almost a half century away (except for the early 1900s rubber boom). In the meantime, much of the literature about the place did nothing to advance our understanding of it, as many titles attest: *Wilderness of Fools; Gold, Diamonds and Orchids; Across the River of Death; Jungle Wife; Green Hell; Lost in the Wilds of Brazil;* and our personal favorite, *Head Hunters of the Amazon* by F. W. Up de Graff.

From the time of Carvajal's initial accounts of bare-breasted female warriors, the Amazon has been shrouded in myth, in part because its vastness, biological complexity, and inhospitable climate have made it difficult for anyone to explore, let alone understand. Over the last twenty-five years, we have traveled through the Peruvian Amazon, touched the Colombian Amazon, and extensively traversed the Brazilian Amazon, and we have met only a few people who've logged more miles in the region. Yet we've never been to Bolivia (other than to touch the soil), Venezuela, Suriname, and Guyana, each of which contains a large portion of Amazon forest within its borders. Pushpins on our personal map of Brazil mark the places we've been; there are none in the state of Roraima (about the size of Utah), and there are gaps of tens of thousands of square miles in the states of Amazonas and Mato Grosso. It is hubris to claim to be an "expert on the Amazon," as well as a near physical impossibility.

The entire Amazon basin spreads for 2.5 million square miles—larger than the size of the continental United States west of the Mississippi. No one could reasonably claim to know the French Quarter in New Orleans after spending a week in Boise, or San Diego after skiing in Aspen; the wildlife in Yosemite Park differs remarkably from that in downtown Minneapolis. That's the breadth of the Amazon.

The river itself starts in the Peruvian Andes and runs 4,007 miles along the equator to the Atlantic Ocean. In *Amazonia Without Myths,* a compendium of purported facts, the authors claim that the Amazon is "6,762 kilometers [4,203 miles] in length, longer than the Nile River system (6,671 kilometers [4,145 miles]) which for some time was considered the longest river in the world." The Nile's still longer, according to most sources, but the point is that the numbers about the Amazon are so daunting that even the "facts" can seem otherwise.

Hardly a reference is made to the river without pointing out its Guinness-like list of trivia, some of which is impossible to verify: Each hour the Amazon River discharges 170 billion gallons of water into the Atlantic, staining the sea brown with silt for 150 miles. The Amazon's daily flow is eleven times the discharge of the Mississippi, providing almost one fifth of the entire daily flow of freshwater into all the world's oceans. (The presence of freshwater aroused the curiosity of Vicente Yáñez Pinzón, who, in 1500, dubbed the mouth of the Amazon the Sweet Sea, unaware that he had actually "discovered" a river.) The river is called the *Rio Mar,* or River Sea, by Brazilians, and oceangoing freighters can journey all the way to the port of Iquitos, Peru, 2,300 miles upriver from the Atlantic. The Amazon basin is drained by two hundred major tributaries; some "fact finders" measure seventeen of them at more than 1,000 miles long, while others claim there are only seven that long. It depends, in part, whether one stops measuring when rivers change names or when they ultimately spill out into the main Amazon.

The river draws its initial strength from hundreds of small streams in the Andes; it tumbles sixteen hundred feet through steep gorges in the first six hundred miles and eventually forms a mile-wide flow in northern Peru. The river then swells up to a two-hundred-mile-wide exit to the Atlantic. As it crosses the Peruvian border, the Amazon runs into an enormous shallow bowl that makes up most of the center of South America. It is a remarkably flat basin that, if stripped of trees, might resemble the Great Plains of the United States or the Sahara of Africa. Through the middle of the bowl, the Amazon River cuts a deep trough that drops only one hundred feet in its final two-thousand-mile course to the Atlantic. The river's immense power comes from many of the large tributaries that rumble into the channel, pushing an estimated 160 million tons of silt a year along the floodplain.

Geologically, the basin sits between two of the oldest rock formations on earth—the Guyana Shield to the north and the Brazilian Shield to the south. These prominent mountain ranges, each perhaps 600 million years old, are both worn down to rippled plateaus. The area between them was most likely an ancient sea bottom that drained during the Carboniferous Period (about 350 million years ago), creating a series of rivers that flowed westward into the Pacific Ocean. The Amazon's source is about one hundred miles from the Pacific, but its ocean exit is four thousand miles away at the Atlantic. Tens of millions of years ago as the Andes began to push up, they bottled up the rivers and created an inland sea. This trapped water eventually broke through the lowlands that connected the two shields near what is now the city of Obidos, Brazil, and cut its way to the Atlantic, grooving into the soft sediment with such force that in some places the river's main channel is more than two hundred feet deep. Over the intervening millions of years, the shape of the basin and the rivers remained the same, though the area has probably extensively flooded and drained several times as the world's oceans have risen and fallen with ice ages.

Even now when the basin, which is mere feet above sea level over much of its area, suffers a particularly bad flood season, the region begins to resemble an inland sea with the distance between banks expanding from five to eighteen miles.

The evolution of the Amazon's weather is still not fully understood, but probably because of the way the Andes block rain clouds from moving west, the area became, on average, one of the wettest places in the world. Whether the Amazon should be called a jungle—"tropical moist forest" is the technical term—is a matter of dispute. The climate varies enormously among different parts of the basin and during different seasons. The maximum rainfall in the western Amazon—the lowland Andes—exceeds 150 inches a year; even in the central Amazon, which has a pronounced dry season and vegetation more reminiscent of a North American hardwood forest, the total is usually well above 60 inches. The temperature averages 75 degrees Fahrenheit throughout the year, though it is often above 90. Humidity, a phenomenon no visitor forgets, hovers at about 80 percent. The rain distribution is roughly divided in half. The forest north of the equator gets heavy rain from April through August, and the southern part gets it from December to April. This variation helps keep the Amazon River floods in check, though the river tends to rise about forty feet each year until it crests in May or June.

Such a wet, stable climate has produced a huge variety of plant life. The vegetation that covers the basin makes up one half of the remaining rain forest on earth. It was once thought that those trees provided more than 80 percent of the world's new oxygen, although more recent studies have consistently disproved that tenet of modern environmentalism. The Amazon is not "the lungs of the earth" but a mature forest. As such, it maintains an equilibrium between the production and absorption of oxygen.

The basin, taken as a whole, is the richest area on the globe in terms of the variety of living things. Perhaps one million of the world's estimated five million species of plants and animals live in the Amazon; many live nowhere else. In 1983, Terry Erwin of the Smithsonian

Institution coined the term "the biotic frontier" in a breakthrough article explaining how we were vastly understating the number of extant
species in the world's rain forests. Working in Panama, Erwin calcu-
lated that at the time there were "163 species of beetles living in the
canopy of a single species of tree. In turn, there are about 50,000
species of tropical trees worldwide, so this comes out to 8,150,000
species of beetle. Then, assuming that beetles represent 40% of all
arthropod species, and finally that there are as many arthropods
(mostly insects) in the canopy than on the ground, [Erwin] came up
with a rounded estimate of 30 million species for the world's rainforests." His research blew past previous estimates by a magnitude of
nearly twenty. Since then, scientists have pared the figure to about five
to six million, but no one will concede that's the limit.

Erwin's work led to a new appreciation for the variation in species
in neighboring tracts in the Amazon. He showed that one hectare in
the forest near Manaus hosted three hundred different types of trees,
and that there was little redundancy in a hectare of forest less than
sixty-two miles away. Each tree hosts a unique family of insects, and
some trees host up to fifteen hundred species, an astounding abundance of different life-forms in a small space. Scientists have also catalogued twenty-five hundred snake species, 2,000 fish (by comparison,
the Mississippi has 250), fifteen hundred birds, and fifty thousand
higher plants—about a fifth of the world's total. Little is known about
most of this life, and scientists estimate that at least an equal number
of species in all categories have yet to be discovered.

In the late 1980s E. O. Wilson of Harvard and Peter Raven of the
Missouri Botanical Garden introduced the term "biodiversity," which
has served as a watchword for the multiplicity and complexity of life-
forms in the rain forest. Most important for the future of the Amazon,
there are now millions of the human species living there: twenty-one
million in the Brazilian Amazon alone.

CHAPTER 3

THE FRONTIER WITHIN

—

In Manaus—now a city of almost two million people, which is three times larger than when we first visited in 1980—a new, more modern shopping mall seems to open every week, pushing the rain forest farther and farther from sight. The newest mall is connected to a first-rate business hotel and a medical office complex and includes three restaurants, two of which are national chains. We sat in the blessed air-conditioning of a São Paulo–based barbecue restaurant with our thriving-entrepreneur friend, Jaime Benchimol, and recounted our latest interviews with loggers and miners in the hinterland. Jaime, born and raised in Manaus, grimaced, as he often did when we referred to the "jungle" and the "wild frontier." He then said, "The problem with what you're writing about is you're chasing after a Brazil that is getting smaller and smaller every day."

He fears that another book on the Amazon will perpetuate a stereotype that Brazilians walk around with fruit on their heads like Carmen Miranda, or that they wear grass skirts and hunt monkeys or each other, or that they live in teeming slums and are stuck in an endless cycle of poverty. These misperceptions bother him as much as Americans' inability to distinguish among Brazil, Bolivia, or Belize, or even to care which country speaks Portuguese, Spanish, or English. When the United States focuses on the south at all, it obsesses over the three C's—Castro, Chavez, and cocaine—a perspective one acquires south of the equator.

Many Brazilians are astonished and saddened that their continent-size country with abundant natural resources and almost 190 million people doesn't seem to matter to the United States, especially when contrasted with Brazil's up-to-date following of news in American politics and culture. Few countries have been as consistently loyal to the United States as Brazil and are as potentially influential throughout South America (only Ecuador and Chile lack borders with Brazil). Yet Brazil remains unknown, even exotic, to Americans.

Brazil, for much of the latter twentieth century, was commonly known as "the country of the future—and always will be," with one foot stuck in the nineteenth century and the other in the twenty-first. It acquired the nickname Belindia to connote its Belgium-like wealth and India-like poverty, and seemed to have a tendency to snatch economic defeat from the jaws of prosperity. Even following eight years of the administration of Fernando Henrique Cardoso from 1995 to 2003, which provided political stability and economic discipline, Brazil's economy dropped from the world's eighth largest to fourteenth by 2004. Just as it has now begun to mature as a democracy and realize its economic potential as an industrial and agricultural powerhouse, along comes a revitalized India and a burgeoning China to steal the spotlight.

Globalization, however, has firmly permeated the Brazilian economy, and this time it's hard to believe the positive economic changes will be as fleeting as some of the past episodes. Notable false starts include the rubber boom in Manaus one hundred years ago, which made the city as wealthy as Saudi Arabia during oil's heyday. Another false start was the Brazilian Miracle of the 1960s, which brought piles of money from international bankers seduced by double-digit annual growth rates—loans that were never repaid, because the 1973 and

1979 oil shocks and balance-of-payments deficits brought about uncontrollable inflation.

What makes this moment different is that Brazil has invested heavily in information technology and the most important asset of the information age: access. Wireless technology has provided a windfall to Brazil, solving the endemic physical difficulty and expense of connecting such a vast country by landlines. Standing near the Bolivian border in the outpost of Rio Branco, or a continent away, nearly 150 miles off the Atlantic coast on the remote island of Fernando de Noronha, the Galápagos of Brazil, we never lost touch with instant access to e-mail or cell phone service. Friends at home, who might have assumed we were wrestling anacondas, found us taking care of business as if we were sitting in our offices. The information available on the Internet, as well as the ability to communicate instantly and inexpensively with anyone, anywhere, belongs to the people of Brazil. The financial capital of Europe and the United States can no longer hoard intellectual capital, the seed for economic innovation. Brazil has joined what Tom Friedman calls "The Flat World." This access—allowing producers to find consumers and consumers to make choices, making research available to farmers, and providing educational opportunities—will change Brazil permanently and positively, as it has for other emerging economies.

The recent discovery of a society in the Amazon contemporaneous with North America's Clovis Man suggests that parallel universes existed in New Mexico and the Amazon, a viewpoint at odds with North American primacy. The United States and Brazil are so much alike in so many ways—nations of immigrants and racial diversity; nations marred by a history of slavery; resource-rich continent-size nations, speaking one language, that remain united despite regional peculiarities—that it would have been reasonable to expect them to have matured in tandem. But the development of the two countries has been far from parallel.

As a consequence of being a colony of Portugal, which avoided the social and political enlightenment of the seventeenth and eighteenth centuries, Brazil lost out on the movement toward participatory government. Portugal was "peopled by an aristocracy exercising power under a hereditary monarchy. Nothing could have been more antiegalitarian." Probably the most significant reason why Brazil did not develop in lockstep with the United States has to do with the distribution of wealth, especially land ownership. Brazil is a country of haves and have-nots, a carryover from the Portuguese system of favoritism and nepotism, or what one scholar characterized as a "patrimonial and paternalistic society." In contrast, the founders of the United States and the generations that followed promoted a system of individual land ownership (with the glaring exception of slavery), which formed the basis of stable government. Universally available state-financed education prepared the U.S. population for the opportunities an egalitarian economy presented. Brazil, by comparison, suffered from limited participation in government, an absence of equity in its legal and social systems, and a population-wide education deficit.

Blame can also be traced to a simple demographic fact: there weren't enough Portuguese to colonize such a large country. To establish rule over so much territory, the royal family divided Brazil into fifteen so-called captaincies, parallel strips of land bordered to the east by the Atlantic Ocean and to the west by the line created by the Treaty of Tordesillas. These captaincies, involving virtual sovereignties of these lands, were subdivided into large territories with interlocking control among friends, family, and other favorites. Landownership was concentrated in just a few hands. Brazil never provided opportunities to its citizens along the lines of the Homestead Act, which granted land to anyone in the United States who would settle it, spreading economic equity to the newer states. And it never had a civil war, as Mexico did, with the resulting dissolution of large landholdings. Even today, more than 50 percent of the arable land in Brazil is owned by just 3 percent of the people. Predictably, wealth follows a similar pattern. The wealthiest 1 percent earns more than the poorest 50 percent. According to *The*

Economist, Brazil is "the world's fourth-largest democracy but third most unequal country."

Land ownership issues underlie the recent past of Amazonia and will define, to a great extent, its future development. The ideal stated by the military government (and disbelieved by many) in the 1970s in the early stages of Amazon migration was that the immigrants would receive clear title, develop a sense of permanence and ownership, and build communities as a source of stability and life enhancement. Settlement of the Amazon, the promise went, would create a vested land-owning middle class, a change from the remnants of the inequitable land system in the rest of Brazil. That never happened.

In the late 1970s and early 1980s, when we accompanied the Amazon immigrants desperate for a plot of land to call their own, we could sense the hunger for homesites. The prospect of bettering one's lot motivated tens of thousands of families to move into untouched rain forest or land claimed by distant tribes. The country wasn't ready then, and really still isn't, to reconcile the needs of immigration with an antiquated titling system, which tilts in favor of those willing and capable of corrupting it. Land wars still occupy the Amazon, creating casualties on a daily basis—some people are evicted from land they may have occupied for years, while others are murdered in cold blood. Many argue that if people actually owned land, instead of laying claim to it with false titles or with force, then the rate of deforestation would fall. Disrespect for the land, the argument goes, arises when there is no investment, of either capital or sweat equity. When land is up for grabs—and possession becomes the law—then destruction becomes the surest display of ownership.

The entire royal family of Portugal fled to Brazil when Napoleon's General Junot arrived in Lisbon in 1808. The family ran Portugal's affairs from Rio de Janeiro until 1821, the only European country to be administered from a colony. Brazil declared its independence from Portugal in 1822, but unlike the United States, which emerged immediately into a democracy, Brazil became an empire ruled by the son of the king of Portugal. A line of heirs to the Portuguese throne governed the country until 1889, when Brazil became a republic.

In his insightful and entertaining book *The Brazilians*, Joseph A. Page points to the early republic as a tipping point for the country for several reasons. Slavery was abolished in 1888. Before that, Brazil had imported 3.5 million slaves, more than six times the number of the United States, because the land was so vast and the native worker pool so sparse. Slavery's end marked the beginning of a great wave of immigration into Brazil as landowners sought a source of cheap labor to replace the freed slaves. Anywhere you go in Brazil, you will meet people with Japanese, Italian, and German surnames, most of whose ancestors arrived during this period, and many of whom still hold on to their native culture as tightly as ethnic groups in the United States do. You can find first-rate pizza in the small Amazon town of Sinop, which was settled by Italian descendants from the south. And there is fresh sushi in Belém because Japanese immigrants established large pepper plantations nearby.

The early twentieth century brought prosperity for Brazil, because of the world's demand for its agricultural products—coffee, sugar, and rubber. But the dislocations brought about because of the collapse of those markets by the 1920s had profound effects on the country. The rural workers fled to the cities in search of work, giving birth to the squalid urban slums (*favelas*) that sprang up in the beginning of the century. The other significant impact from the poor economy was political. Getúlio Vargas, a dictator who ruled the country from 1930 to 1954 (albeit with interruption), rose to power with an odd coalition of support for a social welfare state that would promote both industrialization and unionization. Vargas had a remarkable ability to satisfy both workers (although he opposed Communism) and the ruling class. He created many of the gigantic state enterprises still extant today, such as the oil company Petrobras, and his programs led to government control of nearly all areas of the economy. Private enterprise and entrepreneurship suffered. Capital belonged to the government and its parasitic monopolies.

Vargas committed suicide following a political crisis in 1954. Shortly thereafter, Juscelino Kubitschek, a former governor of the state of Minas Gerais, took office. Known by his initials, JK, Kubitschek injected a new dose of patriotism into the populace, much as JFK did in the United States a few years later. JK's motto was "50 years of progress in 5," and he nearly lived up to it. He created national industries in automobiles, steel, petrochemicals, and large-scale construction, and he protected these industries from competition by, for example, banning the importation of foreign cars. During his administration, steel output doubled, road construction mileage expanded by 80 percent, and hydroelectric power output increased threefold. His most significant achievement, however, was the construction of the capital city of Brasília.

As the child of seafaring Portugal, Brazil settled itself along its strikingly beautiful coast, which provided easy trading connections with Europe and the United States. Consequently, a vast majority of its population lived on a small fraction of available territory. The interior of the country was (and still is) as foreign to most Brazilians as to the rest of the world. There had been forays beyond the settled coast, most notably by adventurers known as *bandeirantes,* who clashed with the missionaries because of their "cavalier disregard of many rules of Christian conduct and . . . insatiable lust for Indian slaves." As Page points out in *The Brazilians,* "these intrepid individuals had two remarkable accomplishments to their credit": expanding Brazil's borders and unearthing the country's mineral riches, the only impetus for settling the interior prior to JK's administration.

By the time JK took office in 1955, the inward move of the country had stalled. Overpopulation along the coast and underutilized land in the interior troubled economic planners. JK proposed, as part of his election platform, that a city be built in the interior, connected to the rest of the country by highways and an air hub, and that the capital be moved there. (The idea had surfaced from time to time and was even incorporated into the Constitution of 1891 as a long-term objective, and the site appeared in schoolbook maps as "the future capital of Brazil.") Those who had grown used to the comfort and sophistication of the capital in Rio, and its proximity to the business hub of São Paulo, fought the plan. But JK, whose grand vision matched the grand dreams of his country, would not be deterred.

The inexorable development of the Amazon in the late twentieth century began with the inauguration of Brasília in 1960. The city would come to symbolize Brazil's ambition and capability, and its location would redirect the focus of the country's physical expansion. Opportunities now lay in the interior, and Brasília was the first step in making that frontier accessible. The construction of the Belém-Brasília Highway, the first road connection between the Amazon and the rest of Brazil, exposed the rain forest to waves of immigration. Prior to that time, all population in the region had been river-based and isolated from the rest of the country. That the Amazon would become integrated into Brazil became not only a possibility but a fait accompli once the first shovelful of dirt was turned at the site of the new capital.

To direct the effort, JK chose Oscar Niemeyer as its architect, who claimed his work was inspired by "the curved and the sensual line, the curve that I see in the Brazilian hills, in the body of a lover, in the clouds in the sky and in the ocean waves." JK chose Lucio Costa as the city planner, a disciple of Le Corbusier, the father of the planned, mechanistic city that had become the trend in postwar Europe. (Houses were to be "machines for living in.") What they created in Brasília looks from an aerial view like an airplane. The triangle of powers—the Presidency, Congress, and the Supreme Court—occupy the cockpit. First class is the dual rows of glass boxes housing the cabinet ministries. A business and commercial district is in the center of the plane. The wings are residential blocks, composed of low apartment houses, schools, and stores that form communities, all designated numerically, as in "I live at SQS 316, BLA, Apt. 102." Running through the wings and fuselage are broad highways that meet in intricate webs of cloverleafs, making traffic signals unnecessary. Sidewalks are rare.

Planning marvels aside, the city needed time to grow into itself. It sits on a high, treeless plain of rich red clay, and in the clear air the oversize cumulus clouds seem almost touchable. As there was plenty of space, the buildings were built apart from one another, creating a feeling of abandonment even in the busiest of hours. The modern architecture made the city visually exciting but aesthetically cut off from the rest of the country—purposefully, the nationalistic planners pointed out, to develop a modern Brazilian school of architecture, distinct from the reminders of colonial Portugal along the settled coast. After seeing Brasília for the first time, the Russian astronaut Yury Gagarin exclaimed, "I hadn't expected to reach Mars so soon."

At first, reluctant bureaucrats would fly to Brasília in the morning and home to São Paulo or Rio at night. The embassies couldn't conduct much business in isolation, so the diplomats followed the commuters. But as time has passed, the city of five hundred thousand people has matured into the cosmopolitan capital that JK envisioned. Unlike Brazil's other cities, however, poor people are scarce. No matter how hard other Brazilian cities may try to hide their *favelas,* they are too plentiful and too conspicuous to overlook. Not in Brasília. Instead, Brasília is surrounded by satellite cities, which one priest referred to as the "rings of misery." The poor are out of sight. And for many years, they have been out of mind.

The law prohibited JK from serving a second term, and the country's progress eroded after he left office. Although he left a legacy of accomplishment, he also left behind a culture of corruption and a scourge of inflation that arose from so much new economic activity at an accelerated pace. His successor, Jânio Quadros, a former governor of São Paulo, claimed bizarrely that "occult forces" impeded his ability to govern, so he resigned from office, reportedly expecting a popular referendum begging him to reconsider. None came. His successor, João Goulart, never stood a chance. When he took steps to nationalize industries, expropriate unused land, and curtail the flight of money out of the country, he alienated the power elite. Goulart was a Castro sympathizer at a time when the United States wielded the Monroe Doctrine (more specifically, the so-called Roosevelt Corollary) to justify intervention. The Brazilian military, with blessings from the United States, rose up, sent Goulart into exile in Uruguay, and would rule the country from 1964 to 1985. The ensuing administrations of generals had a very military attitude about what to do with the Amazon: they set out to conquer it.

CHAPTER 4

NATIONAL SECURITY AND
INTERNATIONAL ENVIRONMENTALISM

—

The empty space scared the military. They had learned slogans in school that summarized a strategy for dealing with this threat: *ocupar para nao entregar,* occupy so as not to surrender, and *a Amazonia e nossa,* Amazonia is ours.

The generals focused on national security, not preservation. In 1964, there were perhaps 250 indigenous tribes thought to be living beyond contact with modern society. (Even today there reportedly are 17.) The military government feared what would happen if these people were left alone. Or if the space were left empty. What would happen if Colombians crossed the border and explained to those Indians that they really were Colombian, not Brazilian? With a tribal rather than national identity, there'd be no reason to resist, especially if there were an incentive—an outboard motor perhaps—to become Colombian. What would happen if these Indians were integrated into a cocaine economy and became pollinating bees for coca leaves? What would happen if they were taught Spanish and not Portuguese? Or what would happen if they were taught Islam instead of Catholicism?

The answer is that not only would they be different from the rest of Brazil, they might also pose a threat. And that would be anathema to the idea of nationhood. Empty space (or what Brazil's military called Amazonia's "natural permeability") threatened the concept of national identity, and the loss of national identity heightened concerns about national security. As diverse a country as Brazil is, when the leaders in Brasília looked at a map with almost seven thousand miles of borders just in the Amazon—with Venezuela, Peru, Colombia, Bolivia, and the other countries—they felt the need to conquer themselves.

The military's wanting control over the nationality of the Indians simply illustrates the threat the military perceived in the vacuum; the policies themselves addressed conquest of the land, not specifically the indigenous groups. Road building was the method of choice. The Belém-Brasília Highway, completed in 1960, quickly achieved permanence on the eastern flank of the Amazon, sparking an immigration movement that delighted the government: Brazilians were occupying empty Brazilian land. The idea for the TransAmazon Highway in 1970, a twenty-five-hundred-mile connection from the Atlantic coast to the Peruvian border, served two purposes: it appeared as a humane solution for the misery brought about by drought in the northeast, by providing an outlet to a wet and supposedly fertile land, and it appeared to open a pathway for a march of Brazilians into the empty space. "Men with no land to land with no men" was the slogan of that road. The government also promoted ownership of large cattle ranches through the use of fiscal incentives, such as allowing people to buy land in the Amazon instead of paying taxes. Other incentives rewarded clearing the rain forest for grazing areas. These were, without exception, unprofitable enterprises, but they established a Brazilian presence in empty space. While the rest of the world complained about the rates of deforestation in the 1960s and '70s, Brazil's generals saw these same statistics as proof of success. What some used to measure destruction, others used to measure integration.

These efforts proceeded at a time when science and social planning had not yet caught up with the ambitions of the presiding generals.

Charles Wagley, who pioneered an Amazonian research program at the University of Florida, lamented in 1975, "Brazil seems to be attempting to change Amazonia more with patriotic spirit than with true scientific planning." These incursions ignored the environmental impact of these movements, and perpetuated a grossly inequitable system of landownership. "The government planned to open Amazonia as a means to counter the concentration of land ownership in rural areas and as a way to attenuate urban squalor," wrote Wagley's students Marianne Schmink and Charles H. Wood in *Contested Frontiers in Amazonia*. "Actual policies led to an even more skewed distribution of land in the northern countryside and only reproduced the urban ecology of metropolitan Brazil on the frontier." Failures outnumbered successes. Deforestation, social conflict, and rural poverty resulted from programs from Brasilía that were imposed on the region.

The military government abdicated in 1985, and national policy became less concerned with occupation for occupation's sake and more concerned with how to respond to complaints from the rest of the world about Brazil's environmental abuse of itself. In the 1980s, Brazil had to tolerate the bombast from countries whose environments had already suffered serious degradation and whose agricultural interests were being propped up by heavy government subsidies. Al Gore opined, "Contrary to what Brazilians think, the Amazon is not their property, it belongs to all of us." Imagine how this statement played in São Paulo. About as well as farmers in Des Moines would have reacted to Mao Tse-tung saying, "My people are starving, and they need free corn from Iowa, which must be seen as an international resource." François Mitterrand put forth a proposal of "relative sovereignty" for the Amazon, which meant Brazil could house the rain forest but couldn't use it without international approval. British prime minister John Major said, "The international environmental campaigns over the Amazon region have left the step of propaganda to begin an operational step, which definitely may include direct military intervention in the region." And Mikhail Gorbachev, the leader of perhaps the single most environmentally destructive country in history, observed, "Brazil should delegate part of its rights over Amazonia to competent international organizations."

With new civilian leadership taking advantage of international goodwill resulting from a peaceful transition from the military, Brazil appeared willing to recognize the substance of the problem, although the government undoubtedly bristled at the intrusion and the patronizing attitudes behind these statements. The president at the time was José Sarney Costa, a senator from the impoverished northeastern state of Maranhão, a landowner, an oligarch, and a friend of the military. Sarney succeeded the popular Tancredo Neves, the former governor from the state of Minas Gerais, who had been elected in 1985 as the first civilian president since Goulart; he died from a postoperative infection before he ever took office. Sarney's background suggested he would pay lip service to environmentalists while continuing heavy subsidies for large cattle ranches and road building in the region.

But Sarney took office at a time when symbolic storms that portended crisis were developing over the Amazon and bringing the rain forest to center stage. In the Amazon's history, no two years were as significant as 1988 and 1989. Those years expanded the debate about the Amazon beyond the conference rooms in Brasília and into the rest of Brazil, as well as beyond the country's borders.

Sarney was seen, even during his administration, as nothing more than a reactionary nationalist who would ignore, or spite, international environmentalism. That really may have been his true nature, but in hindsight, Sarney's handling of the Amazon's environment is less certain. "Nixon goes to China" is an expression that even Brazilians use to provide logic to apparently contradictory government policies, and it's an expression Brazilians apply here to reconcile what they expected from Sarney and what he actually delivered.

We visited Sarney in late 2003 to discuss the interplay between the development of environmental consciousness in Brazil and the traditional concern for national security. We also asked whether he advocated a program of environmentalism in order to protect the Amazon or in order to preempt international participation. An example of Sarney's mixed message could be seen in the *Nossa Natureza* (Our Nature) program. This was a comprehensive package that created both agencies and environmental legislation that, on the one hand, institutionalized environmental protection and, on the other hand, established a green geopolitical defense shield over the country. Beginning with this program, Brazil co-opted the international environmental agenda and blunted many of the sanctimonious statements from international politicians meddling in Brazil's business.

The notion of a mixed motive surprised Sarney. "It is not true that the changes that occurred, occurred in spite of me; they occurred because of me," he told us in an interview in his ceremonial office as senate president. More than thirteen years after leaving the presidency of the Republic, he still held tremendous sway in national politics. He denied that he was the beneficiary of the law of unintended consequences that played out in 1988 and 1989. Rather, he saw himself as the father of a deliberate plan to respond to a crisis.

His long-standing concern for the environment, Sarney claimed, had been waiting only for the right moment to emerge. "It came once the Berlin Wall fell," he told us. "I knew then that the issue of ecology would become the priority issue of the world. And I knew that the attention of the world would be focused on Brazil because of the Amazon."

The Amazon did capture the world's attention for a variety of reasons, partly because international priorities were shifting from cold war rivalries to issues of multilateralism, such as the environment and global trade. The United States had its own selfish reasons for focusing on the Amazon. During the summer of 1988, the hottest in decades, the litany of woes suffered by American farmers grabbed the media's, and then the nation's, attention. Bands played to raise money to save farms from foreclosure. Native Americans resurrected traditional rain dance tunes, and scores of elderly in the big cities died from heat exhaustion.

All the while, fires burned out of control in Amazonia. More than eight million hectares (nearly twenty million acres) of rain forest had burned in 1987, the culmination of what filmmaker Adrian Cowell called "The Decade of Destruction." In the Amazonian state of Roraima, four million hectares, an area larger than Switzerland, burned in less than a month that year. What scientists had been warning us about for years, now appeared in newspapers as fact: releasing vast quantities of carbon into the air might warm up the atmosphere. This long-ignored hypothesis began to receive attention.

The planet was heating up, and to lose the Amazon would be to lose our lung capacity: a global suffocation, the theory went. That idea, while headline-grabbing, would later lose out to science when researchers showed that the Amazon absorbed as much oxygen as it produced. Yet the carbon-releasing potential of a burning jungle raised alarm among scientists nonetheless, and as global warming moved from the science pages to the front pages, concern for the Amazon took hold. A tree's fate in the Amazon would determine a cow's fate in Texas. A far-fetched theory had become gospel.

But the Amazon belonged to Brazil (and, to a lesser extent, Colombia, Peru, Ecuador, and several other countries), and Brazil wouldn't yield sovereignty over 60 percent of its territory to the United States Senate or the *New York Times* editorial page. For better or worse, Brazil has long resisted foreign intervention in its affairs. The 1957 best-selling treatise of Artur César Ferreira Reis, *A Amazônia e a Cobiça Internacional,* still serves as the playbook for this nationalism. Periodically, this brand of paranoid nationalism erupts, such as with the 1967 Hudson Institute plan to create a series of dams and a navigation system akin to the Great Lakes throughout the Amazon. In July 1993, the national Brazilian weekly magazine *Isto É* reported that "ten out of ten" Brazilian officers believed "the gringos want to take over Amazonia." In April 2003, banners strewn across the overpasses of Brasília proclaimed, TODAY IRAQ FOR ITS OIL, TOMORROW THE AMAZON FOR ITS WATER. An ersatz U.S. schoolbook traveled cyberspace in 2003, purporting to be an example of what American students learn. It teaches that eight Amazon countries are incapable of caring for the Amazon, it says the rain forest must be taken from them, and it implies that invasion by America is nigh.

Until Brazil addressed the Amazon's future internally, which did not happen to the world's satisfaction under the military government, the cows in Texas, it seemed, were destined for some hot summers. It was speculated that this apocalypse would melt polar ice caps and flood Manhattan, and this fear provided yet another rationale for the rumor that the United States was drawing up plans to take over the Amazon: to save Manhattan.

Sarney believes he translated the world's concern into internal progress in order to prove to the world that Brazil could take care of itself. "The irony," he scoffed, "is that [the United States] never took care of the environment as you were developing, and the European countries always have polluted the environment."

The debate leading up to the adoption of the new constitution in 1988 (to replace the 1967 version promulgated by the military) provided an example of how Sarney claims he manifested the concern. "It was important that we show internationally that we were aware of these problems and were developing a national consciousness for the environment to address them," he recalled. He remembers that the debate focused on the huge fires set on landholdings of multinational corporations, most of which the federal government had subsidized. (A fire on Volkswagen's ranch was reported to be "bigger than Belgium." Alex Shoumatoff, who has written about the Amazon for thirty years, jokingly corrected the exaggeration in a later report to "bigger than Rhode Island.")

"I always condemned these incentives as giveaways," Sarney told us. "I thought it was wrong for the government to give fiscal incentives in the north for cutting trees while at the same time we were giving fiscal incentives in the south to grow trees. These big companies were just buying land to get incentives, and they provided nothing but destruction in return."

If Sarney's memory is to be trusted, he supported the environmental protections guaranteed by the constitution, the creation of what would become the Ministry of the Environment, and the creation of IBAMA, Brazil's equivalent to the Environmental Protection Agency. "We were lucky that the international concern over the environment coincided with our drafting of the constitution and this legislation, because it allowed us to address these serious issues at the right time," he said.

The Brazilian Constitution of 1988 represents as environmentally enlightened a statement of national intent as the world has ever seen. Whether it was intended to simply appease the international community at the time—a nation without a recent history of democracy might not appreciate the import of a constitution—or was an actual sea change in public thought is still up for debate. The constitution established the "principle that both property rights and the economic order must be consistent with the protection of the environment." It gave all citizens the right to an "ecologically balanced" environment, which the government was required to provide. (Imagine our Bill of Rights providing for free speech, freedom of religion and assembly, *and* the right to breathe clean air.)

The constitution also introduced the concept of a "popular action," which granted private citizens the right to enforce environmental laws. This was perhaps meaningless in practice, but it was another expression of moving authority away from the military and toward the people. The most dramatic step in this direction took place in the constitution's clear mandate to the states and municipalities that political and fiscal responsibility belonged to them to a greater degree than ever before. In framing this constitution, the national government decentralized itself. As Brazil's democracy matured, this new federalism would determine the course of the Amazon's future and balkanize the country as never before. While the national laws were as effective as the available resources and will required to enforce them, shifting power to the states created pockets of influence that shaped environmental policy regardless of Brasília's intentions.

Sarney upheld the three tenets of the military's Amazon policy: sovereignty, security, and development. But the global trend away from

east-west conflict and toward multilateral organizations and global issues redefined these concepts. In order for Brazil to assert sovereignty
over the Amazon, to keep it secure and create its own development
policy, it had to convince the rest of the world of its commitment to the
international environmental agenda, that it was prepared to play a
leadership role. To protect its nationalist goals, Brazil had to embrace
internationalism.

Sarney's path toward an environmentally responsible administration was also determined by forces beyond his control. In December
1988, Chico Mendes, a relatively unknown rubber tapper in the western Amazon state of Acre, was murdered. Seen as an indefatigable organizer of the impoverished and as a troublemaker by ranchers,
Mendes had organized rural workers in *empates,* standoffs on land
used, but not owned, by rubber tappers. The rubber tappers' sworn enemies, the ranchers, in many instances also did not legally own the
land. But the ranchers had the means to corrupt public officials who
"certified" their ownership, and the ranchers hired gunmen who enforced those certifications. When the son of a rancher shot Chico
Mendes in his backyard in the tiny town of Xapuri, near the border
with Bolivia, the gunman did not silence Mendes's voice, as he'd intended. Instead the ranchers unintentionally attached an identifiable
human face to a cause that had previously been advanced primarily by
scientists and foreign environmentalists. A martyr was born.

Even though Mendes had organized and attended meetings on a
national and international level to address the concerns of the landless
tappers, his cause had remained relatively unknown until his death. "I
was surprised at the reaction to his death," Sarney recalled. "I really
had not heard of him. And it happened in such a faraway place, and
we were facing so many big problems in the rest of Brazil. Then
there's this international reaction, and it took us all by surprise. I think
it helped those of us who were trying to protect the environment." Sarney's recent embrace of Mendes's cause conflicts with the memories
of those who found his administration resistant to Mendes's goals. His
revisionism undoubtedly arises from the lessons of elective office, and

Sarney, still powerful today, recognizes a popular cause as well as any other politician.

The Mendes murder had another ramification both in Brazil and internationally. For the first time, an issue emerged from the jungle with a human element. Not only were trees dying, but people were dying, too. Mendes's death did not fit the mold of the indigenous person who died because he was unable to cope with modernity, or the mold of a political dissident rubbed out by a brutal government. Mendes's murder did spawn reflexive "Save the Amazon" appeals, but the response was not so simple. His death reminded the world that there were millions of people living in the Amazon. Any solution to the environmental problems needed to address this human component as well. The extreme solution to the problem—put barbed wire around the Amazon and keep everyone out—may have played well in the think tanks of Washington and London, but it ignored reality.

There was also some sense of unreality following Mendes's death. Robert Redford and Sonia Braga, Brazil's leading film star, sought the film rights to Mendes's life. Sting signed on to the cause and became a highly visible spokesman for the rain forest. Ice cream scions Ben and Jerry discovered Brazil nuts and put them in their ice cream. Xapuri, where Mendes was murdered, became a familiar dateline in newspapers around the world, and there was as much, if not more, outcry when his killer escaped from prison (later to be recaptured) as when Mendes had been killed.

All of these efforts, from the well-meaning to the exploitative, galvanized Brazil and motivated the government to control the unfolding of this international spectacle. Alex Shoumatoff wrote in his account of Mendes's death, "Chico had become almost instantly a symbol not only of the problems of Amazonia and the rain forest, but of the degradation of the environment everywhere."

In Brazil, Sarney responded by pushing through his Our Nature program, which called for (1) the creation of national parks and reserves, (2) the end of fiscal incentives for cattle and other programs that promoted deforestation, and (3) an array of scientific studies of the region. Cynics attacked the government for papering over the problem. Sarney, they said, had been backed into a corner and had to react for his own political survival and Brazil's credibility in the international community. Today, Sarney disagrees. "We had to make a statement in Brazil and internationally that we were creating a national awareness of the environment. This program allowed us to do that," he told us. The program provided the first of many steps that created a populist awareness of the environment, beginning a movement similar to that in the United States in the early 1970s.

The Our Nature program offered, on paper at least, many new areas of governmental involvement in the environment. It recognized the degradation caused by mining and fires, and it highlighted the need for research and the need to take large blocks of land out of circulation and preserve them as government reserves and national forests. Critics pointed out how the program failed to address the pace of large-scale developments and the pressing need for land reform. Nonetheless, the young democracy had created a policy that could be debated in the legislature and the press, and this introduced the vocabulary of environmental protection into the public discourse.

Yet, perhaps Sarney's most significant move was his approving Brazil's surprising bid to host the United Nations Conference on Environment and Development in 1992, which was considered the most important gathering of world leaders, both political and scientific, in the history of environmentalism. Brazil's decision to host the twentieth anniversary of the first gathering on the environment in Stockholm signified a lessening of its fear of international intervention. The decision was also an expression of self-confidence; Brazil understood the magnitude of the problems and intended to lead the way to the solutions. Brazil's senior emissary at the conference had been quoted twenty years earlier at Stockholm asking the developed countries to "give us your pollution" as an invitation to multinationals to industrialize Brazil. By 1992, he stood at the vanguard of Brazil's environmental movement.

"We showed the world that we were addressing these issues, taking the lead on them, and that we could preserve our own resources. No other country with tropical forests could say that," Sarney recalled.

Defenders of the environment around the world cheered Brazil's breakthrough outlook on the environment. As the Rio Summit was being planned, the developed countries met in Houston in 1990 and proposed a comprehensive program in order to oversee the development of the Amazon. Called the Pilot Program to Conserve Brazilian Rain Forests, or PPG-7, it led to the establishment of significant national forest reserves. Yet Brazilian nationalism had not evaporated entirely, and the government's refusal to relinquish its sovereignty over environmental policy provided an impetus for crucial change in Brazilian social policy. Brazil insisted on being a major player in the PPG-7 project, which, in turn, required that it identify qualified participants by training a new generation of scientists, researchers, and social planners. This effort induced the birth in Brazil of a cadre of young people who were to populate international and national organizations and provide a thoroughly Brazilian imprint on environmental oversight of the Amazon. Brazilian-based NGOs quickly developed substantive knowledge and political skills that made Brazil a paramount presence in any discussion of the Amazon's future.

The Sarney administration put Brazil squarely on a path toward integrating the Amazon into the nation's dialogue and policy agenda. The Amazon was no longer seen as a place to be settled primarily by well-connected industrialist absentee landlords and the dispossessed from the rest of the country. In 1955, Kubitschek envisioned that Brazil would look inward toward Brasília; by 1989 that horizon was expanding, as he had hoped. The end of Amazon's isolation would be fraught with promise and peril.

CHAPTER 5

NATURAL WONDERS OF THIS WORLD

—

The promise of the Amazon presents a conundrum of how to view this region and what to advocate for its future. While the Amazon is nature's last great preserve, it also has become, as TR predicted, a great human frontier, unquestionably a land of opportunity. Our random, unscientific surveys in hotel lobbies and restaurants, especially along the BR-364 corridor on the western flank, repeatedly proved that thousands of pioneer families from the south produced a second generation of educated, productive professionals who consider the Amazon their home. They occupy a vastly different Amazonian landscape from the place their parents encountered in the 1970s and 1980s, when they came armed with machetes and matches. The rain forest, even to those who live in its shadow in cities such as Manaus, Belém, Porto Velho, Rio Branco, and Cuiabá, is an alien place.

The revelation that these were environmentally sensitive indigenous groups over widespread areas five hundred and more years ago provides proof of past human occupation. But the discovery does not provide much hope, despite ongoing research, that twenty-first-century occupation will replicate this harmony. Today, there are too many people with too many needs and too many tools. They compete with nature for air, water, and space. If humans win out (as they inevitably will), then nature loses—there is no balance. Any human development—whether it's called rational, sustainable, or environmentally friendly—alters nature's world and interrupts *its* development.

Many argue that this interruption doesn't matter—any alteration of nature that benefits human development is acceptable, since we are a higher life-form and natural selection entitles us to an exploitative existence, especially if we reap rewards. "Kill a tree to feed a child," a slogan heard within the ostensibly "green" Lula administration to justify economic development, fits this model.

On the other hand, destruction of nature still is an act imbued with ignorance, because we lack the ability to understand the ramifications. No one knows for certain the impact of deforestation of the Amazon on regional climates, or deforestation's role in global warming. Nor can we predict the effects of the loss of life-forms that enhance our quality of life. In 2005, the biologist David G. Campbell wrote a lyrical book, *A Land of Ghosts*, lamenting losses of gargantuan chunks of rain forest and the lessons of life within them. "All of this beauty—this refuge of the imagination will end soon," he wrote. "My generation will be the last to live in a species-rich world, in a time when most taxa remain to be discovered. And my generation will watch that world end. Our own species is forging the next great earthly extinction, diminishing forever our only homeland. . . . The changes being wrought in Amazonia will alter the trajectory of life on Earth. The decisions that we make now, at the cusp of two millennia, will have reverberations five hundred years from now, and five million years."

The evolutionary development of the rain forest will never be repeated in a thousand lifetimes. And its deconstruction, cattle ranch by cattle ranch or hydroelectric project by hydroelectric project, causes so many cataclysmic changes so quickly on microscopic to global scales that we can't keep up with its effects.

The rain forest is a fully integrated system where every species connects to another, and understanding these relationships allows us to see how any intrusion, no matter how mild, violates this world. Ghillean Prance, an expert on Amazon botany, has described how the reproduction of the giant Brazil nut tree relies on a specific high-flying bee and a rat. The bee pollinates the flowers in the tree's crown, and the rodent later disperses the seeds that fall to the ground, ensuring that these seeds will not all be consumed by insects and that they will be moved beyond competition from the mother tree. When the forest is flooded, seed-eating fish apparently perform this dispersal function. Candice Millard captures the sense of drama in *The River of Doubt* in her description of TR's immersion in the rain forest, calling it "the greatest natural battlefield anywhere on the planet, hosting an unremitting and remorseless fight for survival that occupied every single one of its inhabitants, every minute of every day. Though frequently impossible for a casual observer to discern, every inch of space was alive—from the black, teeming soil under Roosevelt's boots to the top of the canopy far above his head—and everything was connected."

Nature, which appears so hostile to a solitary person, performs harmoniously by itself even though it seems to be in the midst of a million civil wars. In *Tropical Nature*, an endearing primer on life in the rain forest, Adrian Forsyth and Ken Miyata describe the orderliness of the relationships underlying the apparent chaos. They use the example of a pile of dung to illustrate "one of nature's vast spectacles on an intimate scale, competition for a precious resource at its most intense." The first to arrive are "the tiny dung scarab beetles and metallic ottidid flies. The flies quickly set up for mate acquisition, a contest in which males engage in vigorous charging bouts accompanied by wing-flipping. While this is happening, the little dung scarabs go about their simple business: they plunge in and begin feeding. Many of the females are ripe with eggs ready for laying, and they may bury some eggs as they gobble contentedly away." The authors go on to describe how attacking the pile becomes a "cooperative venture" for some beetles while engendering furious fights among others, both for the prize itself and for the opportunity to impress the females of the species. *Tropical Nature* abounds with scenarios of interdependence among plants and animals, conveying the unmistakable message that tinkering with any part of this machine will reverberate throughout.

Very few visitors actually see or appreciate this beauty. Nature's siege of humanity never abates, and there is no "cooperative venture" between the two. Even a short visit inflicts misery through bug bites, rashes, and intestinal difficulties. At the back of the August 2003 issue of *National Geographic,* which featured an article on remote Indian tribes, the writer Scott Wallace confessed, "I'd already been going to the gym a lot, but nothing in my life prepared me for the deprivation and physical exertion of three months in the wilderness." And he had his own food, equipment, and companionship. This overwhelming discomfit undoubtedly led to some self-fulfilling theorizing among early observers that the place had to have been vacant. Who would want to live here? The moral virtue in Meggers's viewpoint, that humans are incapable of coexisting on a meaningful scale with nature's world, finds ample support in intuition and experience.

The heat and humidity make breathing a workout; the air is suffocating, like a damp blanket. Bugs swarm constantly—bite then flee, then another wave arrives. Some bugs can be seen and heard—whizzing, humming, droning. Others, such as *piumes,* have acquired the descriptive nickname "no-see-ums." Their bites appear as red dots or pinhead droplets of blood. *Tropical Nature*'s chapter titled "Eat Me" focuses on varieties of "semiwild or wild fruits of the tropical forest," although the title also aptly describes the sign that humans evidently carry, which is quite visible to the insect kingdom. The appetites we inspire, the authors explain, can be appreciated by acknowledging what we never see on the jungle floor: dead people. "[In] the tropics you are unlikely ever to stumble across a corpse; such treasure troves of spent life are quickly broken down and returned to the world of the living."

We had our own days when death seemed an attractive alternative. We visited Manaus during the outbreak of an "arbor virus," a generic term used to describe diseases too new to identify or cure. There are some, such as Mayaro fever, Micambo virus, and Guared virus, that have been identified, although still without cures. Our particular ailment sent all nutrients and liquids from our bodies for about ten days.

We escaped the more commonplace diseases mosquitoes carry, such as malaria, dengue, and yellow fever, but we witnessed their devastation in the towns newly carved from rain forest, as if the clearings had created a frenzy of retribution from these pests. In some of the towns along the BR-364 corridor, every family experienced multiple bouts of malaria. (One methodical study showed 90.1 percent of settlers in one new town in the state of Rondônia had episodes of malaria in 1986.) During one of our visits to the Evandro Chagas Institute in Belém to seek advice on a rapidly advancing (and unbearably itchy) leg rash, we saw pictures of the handiwork of flies carrying leishmaniasis, which eats away the nose and palate from the inside over a long period of time, and the beetles carrying Chagas disease, another slow-moving ailment that atrophies the heart and esophagus. The rash didn't seem so bad after that.

Tropical Nature describes "Jerry's Maggot," a botfly that bit one of the authors' graduate students. The insects have "evolved two anal hooks that hold them firmly in their meaty burrow. If you pull gently on the larva, these hooks dig in deeper and bind it tightly to your flesh." That's how "Mark's Maggot" reacted to his less than gentle pulling; the hooks eventually had to be carved from his ankle at Georgetown Hospital. And there also were "Mark's Worms," which would gobble up his meals before he could digest them, an effective strategy for dieting, and depart his body onto his sheets at night. The Georgetown doctors eventually defeated them, too.

No one leaves the Amazon without vivid recollections of ants. There are thousands of different species. Edward O. Wilson noted their omnipresence in *The Diversity of Life:* "In the Amazon rain forest they compose more than 10 percent of the biomass of all animals. This means that if you were to collect, dry out, and weigh every animal in a piece of forest, from monkeys and birds down to mites and round-worms, at least 10 percent would consist of these insects alone. Ants make up almost half of the insect biomass overall and 70 percent of the individual insects found in the treetops." Ants fear little, but there's much to fear about them. Indians use some ants as emergency first aid to bite the skin around a cut; the ants are then killed and a natural suture results. However, if they're allowed to live too long, the cut expands, and uncontrollable bleeding results. Visitors quickly learn ants enjoy warm and moist parts of the anatomy, and salt from sweat is an aphrodisiac. They can't be stomped out; some queen ants give birth to three hundred thousand babies in a week.

The ground sports an array of peril. Caterpillars, "cute though they may be, . . . can cause a caustic rash of almost supernatural pain." Other insects "to avoid include some brightly colored stinkbugs, which spray hot cyanide compounds when they are molested, and large kissing bugs that pack a walloping bite." Roaches, which rank among nature's most adaptive creatures, forage the forest floor with as much rabidity as in our urban apartment houses. Their digestive systems have evolved to the point where they can eat almost anything, anywhere. Spiders tend to occupy the tops of trees, where they have easy access to the small frogs and lizards. There are tarantulas and scorpions here, although their bites are rarely fatal. Hideous-looking Hercules beetles can carry two kilograms on their horns. And headlight beetles glow in the dark, letting their mates know where to find a good time.

The diversity differs not only from place to place and habitat to habitat—water to land—but also level to level. In the 1970s and '80s, biologists Donald Perry and Terry Erwin surveyed the jungle canopy and catalogued an entire new universe of species that was unknown on the ground. Perry described the treetops as "overcrowded lifeboats, their upper surfaces choked with plants clinging to the edge."

At this level, the two thousand species of butterflies in the Amazon participate in the fight for survival with a cunning that belies their beauty. Some of the species incapable of stinging their enemies have learned to protect themselves from predation by making themselves toxic. Birds that eat them die. Remarkably, this knowledge spreads in the jungle, and over time these birds develop other diets. How do birds learn to avoid eating a particular butterfly, spread that word to their contemporaries, then pass that word down through generations? It would be difficult even to study the habits of particular species to try to determine an answer. In May 2002, scientists spotted the golden crown dancer, a bird that hadn't been seen since it was discovered in 1957 by the ornithologist Helmut Sick. Seeing a bird once every forty-five years makes it tough to predict behavioral tendencies.

The butterflies highlight the adaptive abilities of the denizens of the forest: camouflage (caterpillars "duplicate the bud scars and bark texture of the twigs on which they feed"), deception (other caterpillars transform their bodies into the head of a viper, intimidating any would-be predators), mimicry (Bates was the first to notice animals mimicking unpalatable creatures to "reduce their own risk of being eaten"), and evolutionary change (birds' eyesight improving to enable them to distinguish between food and poison).

At night, the jungle's activity erupts, especially the sound effects. One of the reasons the vegetation on the ground is so sparse and that in the canopies it is so dense, is that the sun can't penetrate the tree cover (less than 1 percent of its light "manages to percolate to the forest floor at midday"); the moon has no chance. The darkness is absolute, except for the occasional mucus trail left by a land snail or the glow from a headlight beetle. Howler monkeys wail in a high-pitched tone of strangulation; caimans violently splash along the shoreline in search of small fish. A choir of frogs gurgles; some of the frogs are so poisonous that the toxin on their skin would kill anyone whose open wound came into contact with them. The night also brings out the bats. In the state of Pará in the spring of 2004, less than sixty miles from Belém, thirteen people died after being bitten by bats. And it's cold at night, exacerbated by the dampness.

Snakes, obviously, are to be feared. They tend to be shy, unobtrusive until someone bothers them, and that's a dumb thing to do. The bite of a fer-de-lance, a type of pit viper, reduces life expectancy to two additional hours. Some guidebooks describe the ten-foot bushmaster, decorated by diamond-like splotches, as a snake that "rarely bites," but the rest of the sentence invariably is, "when it does, it's always fatal." The sight of a thirty-foot anaconda, which can swallow a small child whole, gives any visitor second thoughts about carrying on.

The water, of course, carries its own perils, as Redmond O'Hanlon noted in *In Trouble Again:* "the electric eel can only deliver its 640 volts before breakfast; the piranha only rips you to bits if you are already bleeding, and the giant catfish merely has a penchant for taking your feet off at the ankles as you do the crawl." And there's the *candiru,* a small fish fond of urine and willing to swim up the urethra to find its source, at which point it swells up and makes extraction very painful.

Other than snakes and fish, few creatures of size are menacing, which is why the dangers are so insidious. No rhinos, lions, or tigers leap out; large animals have trouble surviving in a dense jungle, where roaming for food is laborious. The diversity limits availability of food sources, as there are few replications of a particular plant in small areas. Jaguars and wild boars live here, but they can't travel in packs (the vegetation is too thick) and are so unfamiliar with animals their size that they, too, leave humans alone unless bothered. Because animals of measurable size are few and far between, scientists still come across new species. In June 2004, a Dutch scientist reported the discovery of a gigantic jungle pig, which looks like a boar but has less fat, no goatee, and no ring of white hairs around its neck. It reportedly doesn't smell as bad either. At about the same time, Rob Wallace of the Wildlife Conservation Society discovered a new species of monkey in the Bolivian Amazon.

Plants perform the same symphony of life, replete with adaptive and protective techniques, and they participate in as diverse an ecological system as exists on the planet. They also serve human needs in a variety of ways—from providing lumber for basic construction to being the source of some of our most important medicines. Early naturalists in Amazonia found curare, and quinine for malaria, and this is where the cure for cancer is supposedly waiting to be found, which is why many argue this place should never be touched.

In his remarkable survey of Amazonia's medicinal cornucopia, *Tales of a Shaman's Apprentice,* Mark Plotkin writes, "We collected enough plants to stock a drugstore: the leaves of the nah-puh-de-ot tree, good for foot cramps; the sap of the tah-mo liana, a treatment for earache; and the sap of the kam-hi-det, a cure for toothache. We found the huge ku-tah-de tree, whose bark was a good treatment for malaria; the sprawling ah-kah-d-mah liana, effective against coughs and colds; and the lithe ah-tuh-ri-mah vine, whose sap could be drunk to cure stomachaches." Plotkin's mentor, Richard Schultes, a Harvard professor, created the field of ethnobotany, which views the jungle as a vast, useful medicine chest, the keys to which are held by its indigenous population. During our own walk through the forest near Xapuri in Acre, our guide Nilson Mendes pointed out treatments for hemorrhoids, hepatitis, malaria, and snakebites; a plant that holds water so you don't die of thirst; a plant that coagulates blood; and a "natural Viagra."

It is possible, of course, for a human being to adapt to this life over time. If a bird can learn which butterflies make it sick, a person can learn how to eat and stay healthy. Judgments on what happens in the rain forest, as elsewhere, depend on perspective and knowledge. Insects that torture us actually provide transportation for life-forms from one part of the jungle to another. Pollinating bees are the taxi drivers of biodiversity, taking flora from one inaccessible place to another. We see menacing insects, while the insects see a life-support system. The Apalai Indians in the state of Amapá once told us about the ghosts that made them sick, ghosts they could not see. We looked skeptical. But they would have done the same if we had told them that germs make us sick; we believe in germs. Same ghosts, different words.

The rain forest doesn't present a clear and present danger to anyone. No one outside the rain forest has been threatened by the plants or animals. This natural world has evolved well without intrusion. The indigenous people who do live there lack the skills to survive elsewhere, and they bother no one beyond the forest. Why not just put a fence around it and leave well enough alone?

CHAPTER 6

VOICES OF EXPERIENCE

—

Every year a chunk of land equivalent to an average-size U.S. state disappears from the Amazon. In the year ending August 2004, 16,236 square miles, about twice the size of Massachusetts, were deforested. According to Conservation International, that represents between 1.1 and 1.4 billion trees of four inches or more in diameter. In the year ending August 2003, 15,283 square miles, more than three times the size of Connecticut, were deforested. In the year ending August 2002, 14,567 square miles, or just less than twice the size of New Jersey, were deforested. The record year for deforestation was the year ending August 1995, when 18,056 square miles were deforested. Just think of driving from Washington, D.C., to New York in August 1994. Uninterrupted forest blanketed all of Maryland, Delaware, and New Jersey. If you made the same drive *one year later,* not a tree would be left.

This deforestation took place during a time of heightened environmentalism in Brazil, during a robust return to democracy when a comprehensive body of laws protecting the Amazon had been enacted and supported by broad enforcement powers. The reaction of the Brazilian government and the nongovernmental organizations to these annual figures can be summarized by the Yogi Berra quote, "It's like déjà vu all over again." The so-called experts annually express "shock and surprise" at the figures. The shock subsides, then reappears the following spring. Fingers point at the culprit du jour—the cattle ranchers in some years, the soy farmers in other years, and the migration of small families clearing homesteads in other years. Loggers, miners, and ranchers get denounced regularly; lately, the natural mortality of the forest brought about by its own climate change has been blamed, a "vicious" cycle of self-choking.

And in response, the government usually sets aside another national park equivalent in size to a small American state. A federal department's budget gets increased by more than one hundred million dollars, at least publicly. A government official sometimes resigns. Nongovernmental organizations use the statistics in their annual pleas for contributions. *The New York Times* writes an editorial reminding Brazil that "the rain forest is not a commodity to be exploited for private gain." *The Economist* chides Brazil for its institutions, which are "weak, poorly co-ordinated and prone to corruption and influence-peddling." Policy makers hold up the deforestation as a sign of a healthy economy; after all, a booming domestic construction industry needs sources of wood; other policy makers speculate that the deforestation is a sign of an ailing economy, requiring clearing for agricultural exports to make up for lagging foreign investment. There was a theory that the deforestation cycle peaked every four years to coincide with the presidential elections and the reluctance of the government to enforce laws, but recent statistics show no real slackening from one year to another.

And, from one year to another, the process repeats itself and the Amazon shrinks.

When we first traveled to the Amazon in 1980, about 3 percent had been deforested. Today, about 20 percent is gone. In 1980, we were told that at the rate of exponential deforestation, the entire forest would be gone in twenty-five years. We went back. A lot of forest stood in 2005, and now we've been told that all of the trees will be gone by 2080. We'd like to be able to make that visit.

Behind these gross statistics lie tragedies of the common. On our first visit, we ate Thanksgiving dinner with the Parakana Indians, who had lost most of their tribe to disease and a general inability to adapt to contact with modern society. Doomed to extinction, they were to be moved from their ancestral land, the only place they knew how to hunt in and inhabit, to make room for the Tucurui Dam. Only one hundred Parakana remain. The loss of a heritage, a language, and these people extends beyond the victims. By losing these cultures we impoverish ourselves. We wipe out thousands of years of a particular way of life, and we cannot replicate these assets and their potential. If we're still discovering species of monkeys and wild boar in the Amazon, it's safe to assume that other species will be burned or flooded in thousands of square miles and will therefore never be discovered. By limiting our biodiversity we are limiting the possibilities of improving our lives.

Tom Lovejoy, who now runs the Heinz Foundation, probably deserves more credit than any other individual for popularizing the cause of the Amazon, both internationally and within Brazil. His initial work focused on traditional academic study, such as bird banding. But when he wrote an article ("Highway to Extinction?") questioning the construction of the TransAmazon Highway, then referred to as "The Pride of Brazil," he put forth the then-novel argument that scientists must sensitize policy makers to the environmental impact of development decisions. He envisioned a sort of intellectual sit-in at the planners' door. No thought had been given to integrating environmental concerns into project planning, and there was no precedent to follow even if a preservation agenda were adopted. As conservation management was a new concept at the time, questions that seem so obvious today needed urgent study then: Where should national parks be located—to focus on rich areas of biodiversity and to ensure their sanctity from invaders? How big should the parks be—to minimize interrupting the ongoing evolution and to avoid political controversy that might accompany any governmental land set-asides? Lovejoy enlisted the support of Brazil's pioneering environmentalist, Paulo Nogueira Neto, to press the military government to consider taking large tracts of land out of circulation.

Affiliated with the World Wildlife Fund in the late 1970s, Lovejoy developed a project outside of Manaus to determine what size a forest needed to be in order to maximize species preservation in that forest. Simply put, if saving one hectare preserves 50 species of flora and fauna, saving two hectares preserves 350 species, and saving three hectares preserves 375 species, you get a good idea of how to maximize preservation in a diminishing rain forest. You also can look at how to manage fragments of less than the ideal minimum size. The exercise, however, is obviously far more complicated than it sounds. All the parts are interconnected: if you lose some insects, then you lose some butterflies that eat them, and you lose the seeds the butterflies carry, and the plants that grow from the seeds, and so on. Understanding the dimensions of these losses and their ramifications takes years.

Lovejoy created the project to last twenty years, although it shows no signs of abating. "I had no idea about the rates of change," he said. "I expected if there were changes, they would happen quickly. Also, we found that some fragments lose their species so quickly it's too short a time for implementing conservation measures. I also had no idea of the importance of what we were doing. The more information we got, the more questions we needed to address."

What alarmed Lovejoy most about the results he saw was the snowballing effect of the loss of forest, or what he characterized as the "disturbing and as yet unanswered question of how much deforestation will trigger an irreversible drying trend." Lovejoy told us that the diminishing rain forest carried with it a "negative synergy" that was "sucking away the buffer" planners assumed could be controlled periodically by human intervention, such as a moratorium on logging. He warned, "Annual deforestation does not consist of trivial increments" that can be deemed acceptable or unacceptable or managed by planners. His fragmentation project demonstrated the interconnectedness of the ecological system, that once trauma is inflicted, its consequences cannot be controlled. "Each year brings the cycle closer to the devastating tipping point," he wrote recently. "Increased drying and fire vulnerability would suggest the tipping point is not far off."

Phil Fearnside is another icon among Amazon scientists. An American, he has spent his thirty-year professional career in Manaus studying the destruction caused by large-scale cattle ranching—an unprofitable enterprise propped up by government subsidies—and uncontrolled population growth ("human carrying capacity," he calls it). Fearnside was an early critic of road building in the rain forest, arguing that providing access without any controls would lead to rampant deforestation. It seems that his career has been about sounding alarms, sadly watching his predictions come true, and then moving on to another field of study. "I really don't look at it in terms of victory or defeat," he said. "You have to operate on the assumption that all of this is subject to human will. You cannot be fatalistic. You just try to save the forest tree by tree."

When we visited Fearnside in Manaus, he seemed tired, not from age (he's as prolific as ever, having authored more than three hundred articles) but from shouting into the wind. A tall, gangly professor educated at the University of Michigan, he expressed apologies for the distance from the reception area to his office at INPA—the National Institute for Research in the Amazon. "They've put me in the basement," he said, not sure if it was because the boxes of his research created a load-bearing problem or, more likely, because "if I'm out of sight, I'm out of mind."

His most recent research addresses global warming, how to use the renewal of the Kyoto Protocol to preserve the Amazon. Fearnside has retooled himself into an expert on the politics of climatology. He condemns the United States for "not believing in global warming, which sets us apart from the rest of the world." But he finds equal fault in the position of the European countries, accusing them of abandoning the cause of tropical deforestation in order to pursue selfish economic interests.

"You would expect that the European NGOs would have pushed for saving tropical rain forests through Kyoto," he explained. "But many of them have been saying they're not worth saving because they're going to disappear anyway. And you couple that with their governments who don't want the United States to buy credits for saving tropical forests, because they want the U.S. to reduce its own fossil fuel emissions in order to make things more expensive to produce." The Europeans, Fearnside reasoned, would rather force the United States to spend millions on pollution control, resulting in higher production costs, instead of channeling those funds to preserving tropical forests. Their desire to seek a competitive advantage trumps their desire to preserve.

He sighed like an academic who had been run over by reality. "Pretty unbelievable, isn't it?"

A younger group of scientists has learned from Fearnside's experience that sweeping antidevelopment condemnations, while fact-based, will be trampled by paramount economic and political forces. Dan Nepstad of the Woods Hole Research Center, who started IPAM, a Belém-based NGO, has targeted microclimates as a culprit that can be addressed on-site, rather than focusing on issues involving international treaties requiring U.S. Senate approval. Nepstad and his colleagues have confirmed Lovejoy's fear that deforestation affects deforestation; that is, by altering the microclimates in the region, the cycle of forest loss accelerates. The researchers have relied on the breakthrough work of a Brazilian scientist, Eneas Salati, who proved in the 1980s that the Amazon forest creates its own climate. Vegetation and climate have codependent cycles—mess with one and you affect the other. Nepstad has advocated large-scale forest conservation because of the "tight coupling" of the forest and the climate: "continued transformation to pasture and crop land could disrupt the rainfall patterns that currently sustain these forests, their biological diversity and agricultural production systems."

In order to translate these ideas into practice, Nepstad, breaking with the mold of research scientists, sought out businessmen and local politicians in order to educate them about the importance of sustainable land use. He asked permission to carry out experiments on local land and promised to share the results of the research and help incorporate them into local land-use planning. Since the occupation is inevitable, he feels, the dialogue may as well be, too. He observed, "I just couldn't see how you could close a frontier where there was such an abundance of resources."

Nepstad has embarked on a path different from the traditional environmental approach of organizations such as Conservation International, whose work in the Amazon includes massive purchases of land in order to take them out of circulation. While Nepstad admires these efforts to quarantine country-size forests, he characterized them as impractical "in light of the available resources." He worries about the effectiveness of this approach toward environmentalism because it focuses primarily on preservation rather than conservation of the land under use. "It takes you out of the debate about how to use the land which is actually being used."

Conservation International's efforts and those akin to them, as well meaning as they are, fall close to the stereotype of an international moral superiority that has raised the hackles of Brazil's nationalists. The structural goal is anathema to sovereignty, no matter how many Brazilians are involved in the organization. If parks the size of Switzerland are owned and/or administered by international NGOs, then Brazil's control has been lessened. If the purpose of the preservation plan is to keep people out, the owners of the land won't have reason to engage in dialogue with federal or state governments, or prevailing private economic interests.

Nepstad is pursuing "qualitative preservation, more so than quantitative," which requires capital for land acquisition. His brand of environmentalism begins with engagement with the "occupying forces."

Three areas concern him most: carbon release, climate change, and water systems, "all of which need governance on the frontier." Advocating responsible development in these areas requires a new type of strategic planning that involves environmentalists as well as private landowners and public authorities. For example, Nepstad points out that the requirement to preserve 80 percent of a parcel of land, a federal law throughout much of the Amazon, is meaningless if the 20 percent that's cut is along the riverbanks and facilitates erosion. "We also are trying to point out areas where cutting the forest would have a greater impact on the regional climate than if it's cut elsewhere, or where it would make sense to reforest certain areas in order to improve that region's climate," he explained.

Slowly, businessmen are beginning to understand that science has a place in halting the degradation of the area and improving the quality of their lands: what's good for the environment may be good for business. "Sure, I wish they would leave it all alone," he admitted. "But that's not going to happen. So, if you recognize that expansion is inevitable, then you just can't sulk, take your toys, and go home. You can still better the landscape."

For many years, Everton Vargas was the foreign ministry's top official for the environment and part of Brazil's team that proposed to the United Nations that it sponsor the Rio Summit in 1992. While some argue that the Rio Summit was a failure in light of the sustained deforestation since the conference, Vargas sees otherwise. He sees the progress that resulted from exposing the developed world to the factors underlying the developing world's needs. After all, he pointed out, the environment is not only an environmental issue. "You have to understand that deforestation is not just about the environment," he said. "Deforestation is an economic issue. It will not be avoided simply by saying, 'Don't cut the trees.' You have to say, 'Here's why you don't have to cut the trees.' "

Vargas said he thought the headlines about the rain forest issue in the early 1990s were helpful in drawing attention to the dilemma Brazil was facing—how to develop an economy and preserve a global resource. That brought valuable international assistance, he said. But he complained that the international community generally misunderstands the problem. He's not sure if that misperception is deliberate; the misperception allows developed countries' leaders to feel morally superior by dictating an environmental policy that they themselves ignored while their countries were developing.

"The 'lungs of the world' myth has been disproved," said Vargas. "The Amazon consumes as much oxygen as it produces. We know that now, but it makes good press. Once people started to realize that, it seemed to diminish some of the enthusiasm, and the complexity of the issue became apparent. So you see the same headlines, but the story now is very different."

There are plenty of reasons to be concerned about the Amazon's effect on the world, but depletion of oxygen doesn't seem to be one of them. Despite a scientific consensus that the Amazon is a mature ecosystem that takes up as much oxygen via decaying plant matter as it produces from new growth, the "lungs of the earth" assertion regularly turns up any time there's an Amazon story in the media—even in outlets as careful as *The Washington Post* and the BBC. *The Guardian* in the UK recently called the Amazon "a vast lung producing 20 percent of the earth's oxygen."

Vargas explained that without a clear and present danger of environmental disaster, the Europeans, especially, realize the Amazon issue might be too complex to address effectively. Vargas believes Europe's motives have changed and that the motives are now based solely on selfish interests. Some Europeans see Brazil more as a threat to their economies than as a threat to the environment, particularly European farmers who benefit from government subsidies that impede imports of soybeans and wheat from countries such as Brazil. For all his diplomacy, Vargas, like many Brazilians, resents what he sees as the hypocrisy of Europeans, who demand that Brazil leave the Amazon undeveloped for the good of the world but who shut off markets that could provide Brazil with much-needed export income.

"Does Greenpeace care that Brazilians are out of work because Europe is off limits to our sugar? No. The view of the international community is that the whole forest should be kept in its virgin state behind barbed wire. They never considered that the forest is not virgin. Many people make their living in the forest. The environment has gotten a lot of attention, but it can overshadow the fact that we have so many other social problems." He went through the litany of third world markers: poverty, urban lawlessness, malnutrition, unequal wealth distribution. An emissary for Brazil's environmental issues, Vargas finds it difficult to convey that the environment is a component of a larger agenda.

"Brazil is like Manchester, England, in the 1880s," he said. "A lot of productivity, a lot of social problems, and a lot of pollution. Would England have closed the factories of Manchester? When you talk about environmental controls, you're talking about a lower standard of living. If the rest of the world thinks we have to control the Amazon, what are they going to do about our standard of living? It's like being in the Louvre and it's filled with people who have nothing to eat. Maybe they'd burn the paintings for heat."

Brazilians, Vargas claims, are fed up with stories in the international press and reports by foreign environmental groups that decry Brazil's agricultural prowess and link it to Amazon destruction. The Agriculture Ministry, citing the "distorted focus" of stories in *The Economist, The Guardian,* and other media, issued a statement that warned: "There are strong indications that many of these articles have as their objective the discrediting of Brazilian agro-business competitiveness." Vargas agrees. "In order to compete with the subsidies given by the U.S. and the E.U. to its farmers, we have to expand the agricultural frontier, which is a major cause of deforestation. We have to expand the frontier to be competitive."

When you scratch the surface of this problem, you find issues that go way beyond the simplistic notion of being either pro-development or pro-environment. The environment won't be safe until it is respected and until the institutions charged with its protection are respected. That respect arises from a citizen's sense of involvement in society, a sense of well-being and productivity. It's difficult to be a diplomat on the environmental issue, because the countries that exert the most pressure about the environment are not willing to acknowledge and address the nuances and complexity of Brazil's plight. Ultimately that understanding is key to any long-term solution. "Other countries just are going to have to trust us to take care of the Amazon. That's the way it'll have to be."

But the real root of all problems, he says, is money. "We need capital. We need it for so many things. You have to understand that once we have capital, it doesn't necessarily mean that we can monitor what is happening in the Amazon and enforce the laws. That's what everyone wants. That's what we want. But in Brazil, there are problems other than deforestation. The environment in the cities is a problem. Health is a problem. Sanitation is a problem. Education is a problem. Security of our borders is a problem. There are limited resources and unlimited problems."

Samuel Benchimol died in 2002 at the age of seventy-eight. When we were new to the region, he took a lot of time to explain to us the importance of a holistic approach to addressing the Amazon's issues. We found him then at the appliance store known as Bemol, which he owned, located near the docks in the warren of narrow streets in downtown Manaus. He never made it onto anyone else's list of indispensable sources. He wasn't a scientist, environmentalist, or anthropologist. He sold refrigerators, televisions, and washing machines. He also distributed propane gas and was a small-time exporter of essential oils. The intellectual set never paid much attention to him. He was a businessman, who went home to read and write about the Amazon every night. He did serve a long and distinguished tenure as a professor at the federal university in Manaus, so though no one may have doubted his wisdom, they may just not have appreciated its breadth.

The Benchimol family arrived in the Amazon in the 1820s as part of an exodus of Jewish families from Morocco. Many descendants of these Sephardic Jews are still prominent businesspeople throughout the region. At one time, most of the important merchants of Santarém were Jewish, and their remains can be seen in a most incongruous cemetery—Hebrew writing in the Amazon jungle. The Bensabe family once controlled commerce in Guajara Mirim, the terminus of the Mad Maria Railroad on the Bolivian border. Benchimol's cousin, Isaac Sabba, once held the title of king of the Amazon (a title that roams haphazardly depending on the au courant industry) because of his vast fortune, which he accumulated from an oil refinery, a giant jute-processing plant, and a lucrative contract with Wrigley's for chicle, the base for chewing gum.

Samuel Benchimol's father was born in the Tapajos River town of Aveiros but gravitated to Manaus in the early twentieth century as rubber became the Amazon's money crop. He was a rubber baron, though "not one of the really rich ones," according to Samuel. He lost everything in the prolonged collapse of the Brazilian rubber industry shortly after his son's birth in 1923. Rubber's price had dropped from £655 sterling per ton and was on its way to £34, where it bottomed out in 1932. The Benchimol family moved to the southwestern Amazon, where the once wealthy father tapped trees. "Those were years of fighting, of poverty, of misery, of sickness—years that brought him and all of us the indelible marks of penury," Benchimol wrote in *Amazonia*.

Yet Samuel Benchimol's father lived to see the family flourish again. Despite extreme poverty, all the Benchimol children went to college, and Samuel helped make the Benchimols one of the wealthiest families in Manaus. By the time of his death, the family had nine Bemol stores in Manaus, anchors in new shopping malls, one in Porto Velho, and several on the drawing board. The blue propane gas containers of Fogas, another family business, are as ubiquitous as Coca Cola throughout the northern Amazon.

As both a successful businessman and an idiosyncratic scholar, Benchimol defied easy labeling. His work consisted largely of compilations of personal memorabilia, raw data, and ruminations on the region. His major tome, *Amazonia,* contains everything from his high school term papers to his college thesis to stream-of-consciousness writing on the future of the Amazon economy. He chronicled the disastrous attempt by the Brazilian government to rekindle the rubber boom during World War II by inducing hundreds of thousands of impoverished northeasterners to migrate. He published a useful summary of census data, detailing the growth of the region from 1970 to 1980, and once when we met him in his office, he was mulling over a pile of statistics on the types of fish caught at random towns along the Amazon. "Someone wants to know how many *tucunares* they caught in Tabatinga last March. I'm sure of that. So many numbers. I'll make something out of them," he told us.

On our recent visits we spent time with Samuel's son Jaime, who runs the family's business operations, and who told us that his father's terminal cancer motivated him to write more furiously than ever. "He felt he still had so much explaining to do," Jaime said, "he would spend his time in the doctor's waiting rooms scribbling notes. Then he would stay up all night putting them together in his last book, *Ecological Zenith and Socio-Economic Nadir,* the good and the bad about the Amazon." Jaime laughed. "The contents of the books were always better than the titles."

In an e-mail correspondence, we asked Jaime how his father would have assessed the persistence of the deforestation, considering the progress that has been made in the last twenty-five years in understanding deforestation's causes and effects.

He responded, "Here are some quick points on my father's work:

1. Samuel Benchimol throughout his work always attempted to convince that the Amazon region was very heterogeneous in its flora, fauna, ecosystems, etc., so that it is dangerous and wrong to have uniform policies for the entire region.

2. He fought against the trend to consider nature as the only concern when defining policies for the region. Both native and urban population ought to be factored in the equation. He was very much in favor of developing the region along the following lines: 'The Amazonian world cannot be isolated or alienated from Brasilian and international development, but it will have to sustain itself along four essential parameters, i.e., it must be economically feasible, ecologically adequate, politically balanced, and socially fair.'

3. If we are to refrain from developing the Amazon for the sake of the planet, then polluting nations ought to pay an international environment tax to compensate for the forfeited opportunity costs. We are currently providing a free service to polluting nations and to the planet in keeping a very large region undeveloped at a very high human cost."

We wrote back with one question:

"We know your father thought it fair that the international community compensate Brazil (or the Amazon region) for its cooperation in land use (or non-use). After all, no one from Manaus tells an Iowa farmer how many hectares he needs to keep as virgin forest. However, how did he respond to the argument that this was a process of selling sovereignty? Nationalists would argue that it sets a bad precedent because it acknowledges international domain over Brazilian territory. How did he balance these two arguments?"

Jaime responded:

"I believe he did not see the issue as a sale of sovereignty, but rather as a due reward for preservation. He would probably answer that poverty and underdevelopment are perhaps the ultimate loss of sovereignty."

CHAPTER 7

THERE'S SOMEONE IN OUR GARDEN

—

To see the discrepancy between the potential policing of Amazon deforestation and the actual law enforcement, we taxied a few minutes from our luxurious Hotel Tropical in Manaus to a clean, well-lighted installation, a high-tech top secret complex guarded by armed uniformed officers on the newly paved road to the Eduardo Gomes International Airport. Built at a cost of $1.4 billion and financed through a loan from the Export-Import Bank of the United States and grant money from Sweden, this facility was originally meant to provide sophisticated air control surveillance over the entire Amazon. The system is known as SIVAM (System for Surveillance of the Amazon) and was built by the American company Raytheon. This meant, of course, that some in Brazil saw the system's implementation as an infiltration by the CIA. SIVAM was conceived at the Rio Summit in 1992, primarily for military purposes, but since then it has provided significant promise for environmental protection. If the military could police the Amazon from the sky, the *ocupar para nao entregar* ("occupy so as not to surrender") strategy, justifying the environmentally destructive efforts at road building and cattle ranching for the sake of establishing a Brazilian presence, could abate. "Monitor so as not to surrender" is an ecologically friendly geopolitical alternative. Aerial surveillance could allow the government to discourage intrusions into strategically important, albeit environmentally sensitive, areas. Surveillance could also detect trespassers in reserve areas and allow prompt action. Civilian scientists also appreciated the system's potential to monitor deforestation and climate patterns, and grafted their own program—SIPAM (System for Protection of the Amazon)—onto it.

The combined programs have nine hundred monitoring posts on the ground, nineteen radar stations, five early warning jets, and three remote sensing aircraft all reporting to satellites that transmit information to centers in Manaus, Belém, and Porto Velho. When SIVAM became operational, *The New York Times* reported, "[The] system is so sophisticated and comprehensive that Brazilian officials now boast they can hear a twig snap anywhere in the Amazon."

Young scientists working at SIPAM's headquarters sit on the edge of their chairs as they survey thousands of square miles in a day. Their voyeurism is incandescent. "We can watch illegal logging as it's happening," Luciano Valentim told us. "We can watch people mining gold, and we can confirm what they are doing by monitoring the water samples from the nearby area for traces of mercury." Their ability to monitor activities on the ground has grown so sophisticated that they can detect the probability of selective logging of contraband trees in a densely vegetal region. Valentim showed off detailed satellite maps of the *Terra do Meio* region of Pará, where there had been a persistent problem with the loss of mahogany trees. Pointing to a gouged-out area on the map, he said, "They don't know we're watching them, but we know exactly what they're doing and where they're doing it."

They can monitor forest fires, spy on illegal landing strips, and detect incursions into indigenous lands. There's a hitch, of course. Governor Eduardo Braga of Amazonas told us, "SIVAM/SIPAM can give us the information, but without a ground attack, we can't stop it. We can know about it from the air, but we can only stop it on the ground."

In an ideal world, Valentim could call in IBAMA, the environmental protection agency, to shut down the illegal logging, or he could summon the police to protect the Indian lands. "But we don't have the resources," he lamented. "We can make the calls, but if we don't have the resources to enforce what we have found, then we just end up watching people breaking the law and getting away with it. It's very frustrating."

In November 2003, the call went out to IBAMA and the federal police to shut down illegal logging operations around the TransAmazon Highway town of Medicilandia. But this incident showed how problematic this process can be. The system worked in that SIVAM saw the illegal activity. The system worked in that IBAMA was alerted (through a joint effort with Greenpeace). The system worked in that IBAMA and police officials showed up to stop the loggers. But then the Amazon intruded. The mayor of Medicilandia, himself a logger, rounded up his allies, and they surrounded the government officials in their hotel and held them hostage. They refused to release them until IBAMA dropped its plans for an extractivist reserve in the area, which would have impeded logging operations. (An extractive reserve gives rubber tappers a license to tap trees without ownership rights over the land.)

Monitoring the Amazon and managing the Amazon are activities that should, but often do not, intersect—not only because of scarce resources and rebellious mayors, but also because of the Amazon's vastness. Indigenous reserves work great on paper, for example, but they work less well in the real world, where they represent opportunity to land-starved settlers or mineral-hungry miners. These are not walled-off areas, and they are only as secure as government surveillance and presence can make them. Unless someone answers the phone when SIVAM calls and says, "We're seeing trespassers on Indian land!" then these modern tools are only expensive toys. The clash of civilizations, especially over scarce resources, plays out in the dark corners of this remote rain forest, as violently as on the oil fields of Iraq. In March 2002, for example, the federal police evicted thirteen hundred prospectors from the Roosevelt Indian reservations in the state of Rondônia. The reservations are home to the Cinta Larga tribe, which was first contacted by non-Indians in the early 1960s. This reservation had a history of violent incursions, notably the "Parallel 11 Massacre" in 1963. Gunmen allegedly hired by rubber plantation owners machine-gunned a Cinta Larga village located on the Aripuanã River, killing thirty-seven hundred of the five thousand Indians. SIVAM was meant to detect these incursions and alert authorities before any armed conflict commenced. Still, some areas are so remote, and the government's resources are so strained, that what looks good on paper often fails in the field.

The prospectors returned. In April 2004, after sustained incursions at the largest diamond mine in South America, a war broke out on Cinta Larga land. Twenty-nine prospectors were killed at short range by clubs, spears, and guns.

The press reports revealed the difficulty of investigating these crimes. On April 22, 2004, the Associated Press reported that the reserve where the murders occurred covered "6.7 million acres." The following day, Reuters reported that the reserve was "5.2 million acres." A misunderstanding about the size of Delaware. There were also conflicting reports of where the true blame lay, in part because of an unspoken antipathy toward the Indians that exists across the Amazon. Resentment runs particularly high in the Amazon state of Roraima where so few people control so much land. (In this case, thirteen hundred Cinta Largas live in an area at least three times the size of Delaware.) According to some reports, the tribe's chief had received approximately ten thousand reais from miners for the use of the land. He'd assured the miners that the money would be distributed among the tribe. It apparently had not been. The Indians who attacked thought they were attacking trespassers; the victims thought they had paid for the mining rights.

The security of the Amazon rests disproportionately on the shoulders of one man who seems to show up wherever there's trouble, including in the investigation of the Cinta Larga killings and the expulsion of those miners. We visited him at his office in a complex of one-story white buildings behind a locked gate on the main street of Tabatinga, where Brazil, Colombia, and Peru meet at the Amazon River.

In the office, air conditioners hum and men on cell phones speak in quiet whispers, making secret plans. Computer terminals circle the large room, displaying satellite images similar to those we had seen at SIVAM. Printouts of the images cover the walls, as do radar maps and topographical maps.

Mauro Sposito, the officer in charge, is in his mid-fifties, with thinning gray hair and a military trim. His gaze is intense but his demeanor relaxed. He addresses his individual subordinates as "My love" or "Beautiful." (In Brazil, one must be secure in his manhood to address a federal cop like this.) They ignore his flattery but treat him reverently. Sposito may be the most traveled man in the Amazon. Whenever and wherever trouble has erupted in the Amazon over the last twenty-five years, Mauro Sposito has been there.

As chief of the federal police district in Maraba in the early 1980s, he succeeded in providing some semblance of law and order. There are still shoot-outs concerning land in remote areas, but now the violence in the established towns of Maraba, Xinguara, Rio Maria, and Conceicao da Araguia is simply called street crime, thanks in part to his efforts. In September 2000, Brazil chose him to head Operation COBRA, which is nominally a joint venture between Colombia and Brazil but is really an operation to make sure that Colombia's drug and political problems stop at the border. In 2003, the French government sent intelligence officers to Manaus in an attempt to negotiate with the Colombian kidnappers of a French (and Colombian) national. The French concealed the purpose of the mission. When Brazil learned of the ruse, they called upon Sposito to track down and apprehend the French negotiator.

One experience still haunts him. When Chico Mendes was murdered, Sposito was in charge of the federal police in Acre. He has been accused of failing to protect Mendes after receiving a tip that the assassination was about to occur. Sposito has denied this charge repeatedly, pointing out that there was a policeman with Mendes when he was killed. Mendes, however, frequently had complained that Sposito wanted him dead. Sposito has filed a still-pending defamation lawsuit. Several times he assured us that the charges were false, even when we didn't raise the subject.

None of this history would be noteworthy except for the irony that Sposito himself has achieved heroic status, although not on the scale of Mendes. He has single-handedly created a strategy for bollixing illicit drug production in the outer reaches of the Brazilian Amazon, and he has provided the day-to-day policing against the incursion of the Revolutionary Armed Forces of Colombia (FARC). These armed forces make up one of the most ruthless and resourceful guerrilla organizations in the world, and they operate almost as an independent country in the Colombian territory bordering Brazil.

Sposito rarely leaves his compound without a heavily armed colleague at his side. All that he has accomplished can be dissembled with a carefully placed *narcotraficante* bullet to his head. He has no obvious successor.

"I am alive and I am free," Sposito told us as the driver opened the gate for our visit to Colombia. "You will now see the residences of many people, and they are not alive or they are not free. Until this moment, I am winning," he said, and laughed.

We drove by the Tabatinga airport, where Sposito keeps two of his trophies—the rusted hulks of a DC-3 from Aviopacifico, an Ecuadoran airline, and a twin-engine Carajas plane, gutted by the weather. No other airport would tolerate such junk; here they are monuments.

"The big plane belonged to Evaristo Porras Ardila. You know him?" Sposito finds it inconceivable that we do not. He circled the planes for what must have been the thousandth time. "He was one of the three most important founders of narco-trafficking. He was like a king here. Was."

The king had two colleagues: Roberto Suarez ("The Godfather") and Luis Cardenas Guzman ("Crazy Fly"). Sposito sneered. "They, also, were called the kings of cocaine." In the 1970s the three kings controlled the world's supply of coca paste from Bolivia and Peru, sent to Colombia for refining and export.

Suarez and Guzman would come and go from the Amazon, delivering product and making deals. Porras Ardila set up shop in Leticia, down the street from Tabatinga, and he became the largest employer in the region. The cartel families of Cali and Medellín received world notoriety, but Porras Ardila was the major supplier of drugs to these families. Sposito said he became so rich "he bought a dictator and a country"—Manuel Noriega, whose Panamanian banks served as the financial center of the drug trade.

During our visit, Sposito directed the driver to the hotel in Leticia that literally served as the boardroom for the illicit drug traffic in the 1980s. Ignorant and naïve to the underworld we'd stepped into during our visit in 1981, we stayed at the Anaconda Hotel until an army raid on a pool hall resulted in our being ordered to leave the country the next day. No one could figure out why we were there. No one believed we were there to "research a book." To buy drugs or to arrest drug dealers were the only two reasons people traveled to Leticia.

Porras Ardila began to lose influence when he was charged with planning the murder of Colombia's Minister of Justice, Rodrigo Lara Bonilla, in 1984, a crime that galvanized opposition to drug traffickers and brought the United States into the fight against them. In 1987, Porras Ardila was arrested and charged with ordering the death of a Colombian newspaper editor, and by the end of the decade, Noriega had been ousted. Having lost his patron, Porras Ardila also lost his influence.

The Medellín and Cali cartels took over the vertical operations, from growing leaf to processing, but soon they faced the FARC, which was intent on breaking their paste monopoly. The U.S. success against the coca growers in Bolivia and Peru pushed coca plantations into rural areas of Colombia, where there was no government presence. The FARC supported local growers, promised them a market for their product, and made supply deals with the cartels, which built the refineries and arranged export. Leticia still provided a gateway for the drugs coming from Bolivia by air and Peru by river, and it served as the shipping point for cocaine leaving Colombia.

"I really don't like to go to Leticia, because I have arrested so many people, and their relatives don't like me," Sposito said as we entered Colombia. "At first it was just the families, and they really didn't want to make problems with the police. But now with the FARC, they will kill anyone."

"There's one now," Sposito said. He pointed to a teenager no older than sixteen; the boy was dark skinned and wore jeans, no shirt, rubber boots, and a yellow towel as a bandanna—the rebel uniform. "Just last week a boy of twelve years old was killed. He was in the FARC. They don't care how old you are, or how young." The random violence of the FARC made Sposito yearn for the days of Porras Ardila, where a single power source controlled the flow of drugs. "It's much easier when you can see the enemy."

We drove through Tabatinga and then Leticia, observing scores of teenagers loitering on the dusty streets, dreaming of becoming a soccer star, or getting a job, or a girl—which is hard to do without money. Drug-running recruiters show up with guns and promises of four hundred dollars per month. "My job," said Sposito, "is to make sure what goes on in Colombia stays in Colombia. The FARC know that there is nothing I can do to them in their country, but when they come to Brazil . . ." He made a shooting gesture with his right hand and pulled the trigger.

The economy of Leticia was built on drugs, and Tabatinga was built on the economy of Leticia. "Look at the people who live there," he said, pointing to the muddy sewage-infested shacks along the river in Leticia. "A little money, and they work for the *narcotraficantes*." And a lot of money was once there. Leticia was a city of no more than twenty-five thousand residents, but it had one commercial passenger flight a day with passengers and cargo to Bogotá, and one flight a day with cargo only, an extraordinary amount of traffic for a town that small.

Sposito pointed out the shuttered house of Vicente Wilson Rivera

Gonzalez, another kingpin of the paste trade. It was not the mansion we'd expected but a rutted stucco three-bedroom house behind a rusted gate. According to Sposito, none of these drug kings met a happy ending; they have all been murdered, imprisoned, or impoverished. "That does not stop the others from trying," he warns. "What else do they have?"

He said it was easy to be against illegal drugs sitting in air-conditioned offices in Washington. "But look at this," he said, pointing to garbage in the street and children milling about. It's difficult to sell people on the evils of cocaine when raw sewage passes ceaselessly in front of their houses, or their siblings are dying of dysentery, or they spend their days idly without the prospect of work, baking in the mosquito-infested heat. "If you don't give people the hope that you want them to have, they'll grab the hope others have for them," he said.

As control over drugs passed from the well-managed cartels to the protection rackets of the FARC, a vacuum of authority opened in Leticia and Tabatinga. At the turn of this century, murders take place in both cities at a rate and with the brutality one would expect to see in major urban centers. Sposito keeps a list of the well-known dealers, where they were murdered, and the circumstances. "But when one goes, two arrive," he said.

Sposito sees himself as a doctor with a single hypodermic needle fighting an epidemic. "They gave me one hundred eighty guys," he said. "That's not even what they have guarding a shopping mall in Manaus." He also has eighteen patrol boats, two airplanes, and a helicopter to fight against a rebel movement that is waging civil war against the Colombian army.

His plan was modest. To keep Brazil from becoming Colombia meant keeping Colombians out of Brazil, a daunting task considering the border is 1,021 miles long, all of it untamed wilderness. "In the first year, we found nine airfields being used to transport drugs. We destroyed every one of them." His finger traced the area of the Brazil-Colombia border of the Amazon, an area known as the Dog's Head because of its shape, an area of total lawlessness because of its remoteness from any settlements. When Sposito arrived in 2000, the monthly average of cocaine seizures was 1,000 kilos; three years later, it dropped to 250 kilos.

Sposito explained how the drug trade was working. "In Itaituba," he started, "they have the largest collection of private planes in all of the country." These were left over from the gold boom of the 1980s and '90s. "The pilots fly from Itaituba to small strips in the forest along the border, with drugs, and then go south to Rio, drop the drugs, continue to Paraguay, and pick up ammunition and make the circle again. If a pilot flies the whole trip, he can make two hundred thousand dollars. And he starts again." He whistled. "Rich like gringos," he laughed.

When we saw Sposito, he was preparing to travel to São Gabriel da Cachoeira, a trip of nearly two hours in a twin-engine Carajas airplane from Tabatinga, over land that time forgot. The trip's purpose was to plan the bombing of an airfield that was 174 miles from São Gabriel, just 2 miles from the Colombian border. First built about twenty years ago by a mining company, the airfield was abandoned until the FARC found it and began to use it. The air force had bombed it in Sposito's initial campaign of airfield destruction, but local Indians informed Sposito's men in the area that the FARC had returned to rebuild it, intelligence that the satellites confirmed.

"We're going to make it our Baghdad," he explained. "We will bomb it and then bomb it again." Unlike in conventional war, however, Sposito didn't want the element of surprise. He had informed the press so they could observe the attack. He wanted to make sure that everyone in the region knew well in advance that the airfield would be bombed. "I am not worried that someone will escape with it," he explained. "But I am interested that everyone knows that we know what is happening there. I don't want anyone to get hurt, so I want to be certain that they know the day and time we will bomb. And I want the people to know that if there is any activity to help the FARC, we will know about it."

In 1998, Colombia's army actually entered Brazil in retreat from a defeat by the FARC. Incursions from the FARC, despite the efforts of Sposito and his colleagues in the army, are a daily occurrence. There is no prohibition about legal products going into Colombia, and Brazilians reportedly supply the FARC. Impoverished residents living in border areas without electricity, sewage treatment, or any education cannot resist offers of money to provide food and other supplies across the border.

"We can only stop what we can," Sposito said as he drank a cold beer in a roadside café after working out the logistics for the bombing mission.

Sposito asked us to travel with him to São Gabriel da Cachoeira to observe the final preparations for the bombing. São Gabriel was built by the Portuguese 250 years ago at the rapids of the Rio Negro. The settlement was started as a bulwark against Spanish incursions down the river toward the Amazon, and it was one of the few towns we saw that had natural beauty. Rushing black waters contrast with white sandy beaches. Slender hills stick up like green thumbs in the jungle landscape. Other than the hills around Painted Rock Cave in Monte Alegre, we hadn't seen any elevations in the Amazon.

São Gabriel still has the fortifications the Portuguese built to protect the town against the Spaniards who came down from Ecuador. Why anyone would want to conquer the place is difficult to fathom. The isolated upper reaches of the Rio Negro are significant only because of their isolation. The most meaningful efforts to preserve the virgin forest will take place there. No roads connect the town to any other settlements. You come by boat from Manaus, which takes more than a week, or you come by small plane. It's one of the areas that could be fenced off without affecting commerce or social development. The only conceivable reason to be there is to make sure someone else isn't.

Geraldo Castro, a thirty-one-year-old member of Sposito's team, had just returned to São Gabriel from thirty days in Icana. São Gabriel was city enough for him. "I lived there in Icana with all of the people in the town—three Indian families, twenty people. We watch the river day and night. We look for arms, drugs, gasoline. We stop every boat that comes into Brazil." Such work tests the spirit of even the most dedicated civil servants, but there is no other way to guard the entrances into Brazil. "It's very hot and very boring," Castro concedes. "But it's an important mission for the country." An engineer from Belo Horizonte with a sense of national service—that may be the most important weapon Brazil has.

CHAPTER 8

THE LEGACY OF EL DORADO

—

A fair amount of El Dorado fever still permeates Amazonia, rapacity that results in overnight deforestation, intensive occupation, and then, several years later, abandonment. Cadres of miners carry out the invasions, chasing rumors of gold like itinerant rock-and-roll fans in search of the next festival. Miners combine, scatter, and recombine in different groups, usually finding each other again at a new site years later. Their number fluctuates "as rainfall changes, the agricultural calendar shifts, and the urban economy waxes and wanes," writes David Cleary in *Anatomy of the Amazon Gold Rush*. Cleary chooses three hundred thousand as the best estimate of permanent miners in the region. The Portuguese word for these mines is *"garimpos,"* and the miners themselves are called *"garimpeiros."* They live a hard life, far from any towns and the amenities of electricity and plumbing, and they are constantly confronted by disappointment, which they evidently reconcile because they never leave this life. They have a careless optimism, like a resilient teenager living with disappointment he knows can't kill him.

One of our first visits to the Amazon coincided with the discovery of the Serra Pelada gold mine, a discovery that brought Brazil tangible riches but also served invaluable symbolic purposes. The mine's discovery came to define a transformation in Brazil's society. About sixty-two miles from the regional hub of Maraba, on a small hill in the untouched jungle, distinctive only because few trees grew on it (hence its name "Naked Mountain"), a giant tree supposedly fell over during a rainstorm in January 1980 and exposed a mother lode of gold. In the years since we received this firsthand account, competing reports have surfaced. In his book, Cleary concludes, "The exact circumstances of the discovery are already encrusted by legend, and there are various versions." One legend has it that a miner panning a stream discovered gold rocks and realized that he was standing on riches; another version is that the landowner contacted a geologist after hearing of gold strikes nearby. That soil, everyone agrees, contained rocks of gold. Jungle mines typically produce only shiny specks of metal, even after arduous hours of panning and washing. Within a month, twenty thousand prospectors swarmed over Serra Pelada like ants on a sugar bowl. Most of them had lived a lifetime of toil fueled by dreams of El Dorado, a legend that still lingered five hundred years later. With each shovelful, some miners unearthed more wealth than they had ever believed existed.

José Maria da Silva, a thirty-four-year-old prospector who had mined for gold throughout the Amazon for his entire life without finding more than a day's wage, found 327 kilos of gold in a single day: six million dollars. Over a span of several months, he pulled out 1,300 kilos. Soon after, he divorced his wife. One of the miners told us, "Then he bought a new house and asked a nightclub singer to live with him. He wanted to *dormir com ela* before, but she said no because he was poor. Then he was living with her, then he stopped. Now *ele come* so many girls, young girls. So many."

José Maria, a folk hero to these men, many of whom had never earned more than a subsistence wage before the mines, vowed to us, "I am going to take care of my friends." He promised to build "the greatest motel in the Amazon so that all of my friends can bring their girls there and *comer* like gentlemen with whiskey and air-conditioning.

A clean place." He made good on his promise. The Golden Motel in Maraba is still there twenty-five years later, yet more as a monument to an age than a functioning tryst lair. José Maria lives down the street and tends to the ranches he bought with his good fortune. The Serra Pelada mine has long been closed. He is one of the few miners, we were told, who managed to hold on to his take.

Antonio Gomes Souza, who was forty-four when he came to Serra Pelada, was a friend of José Maria. Souza, his teeth lined with so much gold that his mouth looked like needlepoint art, told us he had found seventeen million cruzeiros (the currency at the time) in gold, about $275,000. The main square in Serra Pelada was called the Plaza of Lies, as instant prosperity begat exaggeration. It was a hive of braggadocio. We asked Gomes to prove his net worth; we had a hard time believing those numbers.

Gomes produced a receipt from the Bank of Brazil. We examined it. The miner was wrong. The bank receipt was not for 17 million *cruzeiros* but for 173 million, about $2.75 million.

Before the mine, Gomes had been a laborer, picking Brazil nuts. "Now I'm going to buy big cattle ranches," he said. He told us of another miner friend who also was very wealthy and who was going to buy five airplanes. "He's crazy." Gomes laughed when he told the story. "He doesn't even know how to drive a car."

Joaquin Almeida, a thirty-two-year-old psychologist from Rio, was watching the news one evening with his wife and three children and saw a report of the Serra Pelada strike. A wakeful night followed. By morning his bags were packed; he kissed his family goodbye and headed off to the jungle to seek his own pot of gold. "I took a bus to Maraba, then got a ride, then I walked the last fourteen hours through the jungle," he told us.

"Were you scared?"

"*Sim, Senhor*," he answered. "I had barely been out of Rio in my life, and here I was in the Amazon jungle. I saw two men dead on my way in, but I think they must have been on their way out and were hijacked."

"Has it been worth it?" we asked.

"Life here is terrible. I miss my wife and my children. I haven't left this place in five months. My house is some black plastic held up by sticks, and when it rains, it falls down. I'm always picking bugs off myself. Sure, it was worth it," he said, winking and rubbing his thumb and index finger together.

Almeida had been making $240 a month in Rio; his wife, also a psychologist, made the same. At Serra Pelada, he cleared $300 *a day* after paying his partner and the sixteen men who carried dirt for him and searched it for gold.

In the early 1980s, these stories reverberated throughout the country with a mania akin to a stock market bubble. El Dorado had been found. Within six months, thirty thousand people flooded into Serra Pelada. Cleary writes about the "economic dislocation" that Serra Pelada caused in the region. Businesses had to close as their blue-collar workers, clerks, and laborers abandoned their jobs to search for gold. Most able-bodied men, bitten by the fever, ran as fast as they could through the jungle to El Dorado. Brazil, a country that had long believed in its natural richness, despite the elusiveness of this richness, had now found it in abundance. And it was not just the gold that excited the country. Geologists continued to expand the reserve estimates of iron ore in the nearby Carajas hills; the bauxite mine in Trombetas just north of the Amazon River near Santarém was coming online and producing jobs and export revenues; the world's third-largest kaolin (the white clay that makes the gloss on magazine paper and the adhesive in antidiarrheal medicine) mine had recently been discovered along the Jari River north of Belém.

Serra Pelada, though, was in many ways the most important of these finds. It was the only one the people were allowed to touch and hold on to. The government or large private enterprises controlled most of the mineral riches in the country. Not at Serra Pelada. Press reports featured a new millionaire every day. And the lucky lot fairly reflected Brazilian society—the destitute, the adventurer, the calculating professional. They came from industrialized cities, from small Amazon towns, and from the other *garimpos* throughout the region. Once there, they lived together in rough makeshift conditions, sleeping under tarps, working through the night, never bathing. The scene dominated media at the time. Our own story appeared on the cover of *Parade* magazine. The photographs of the place that accompanied the text looked like Hieronymus Bosch's *Garden of Earthly Delights*. The Brazilian photographer Sebastião Salgado garnered international awards for his images, which were compared to Dante's *Inferno*.

Miners hacked away at the plots with pickaxes and shovels. Each square plot had been dug to a different level and was connected through a series of handmade ladders. Hoses snaked around and through them. Thousands of mud-covered men, shoulder to shoulder in perpetual motion, passed sacks of dirt from one plot to the next, finally to those bent double carrying them to sluices, where they were filtered.

At first, there was talk of the government's evicting all the miners, declaring the area a national security installation, and bringing in the government-owned mining company. Waste resulting from primitive mining methods was a justification; another was that all this wealth would be mined and no one would report the income on taxes. The government would lose a valuable resource and get nothing in return. Also, the violence in the first months was horrific; fights, fueled by an excess of alcohol, broke out over plot ownership and the affections of scores of prostitutes who flocked to the area. Without sanitation, disease spread among so many in close quarters. Exploitation thrived, too, as imported goods were sold at extortionist rates. These precarious conditions came about in a region of the Amazon with a particular history of social and political unrest.

At that time, Brazil still had a military dictatorship, a controlling central government, and a massive foreign debt. Intervention in the cause of establishing order and taking the national patrimony from those who were taking it for themselves would be consistent with the military's prior policies of social and economic stability. But that never happened. The government's reaction to this phenomenon reflected, in a microcosm, the changes that ended up sweeping the country over the next generation.

In May 1980, Major Curio, a mustachioed army officer with unshorn dark hair and a trim figure, alighted from a helicopter, walked to the center of the rugged dirt runway at Serra Pelada, pulled out his gun, aimed it toward the sky, and pulled the trigger. That act changed the world for tens of thousands of men. For the better.

"While you can stay armed, don't forget that the gun that shoots the loudest is mine," he reportedly shouted to the throng. Within a month, thousands of guns had been laid at his feet. He issued a no-woman/no-alcohol/no-gambling edict, broke up large landholdings that some successful miners had assembled, and required attendance at a morning flag-raising ceremony. He gave everyone already at the site a prospector's work card and promised to bar any newcomers. Plots left unmanned for seventy-two hours were redistributed. He repeatedly tossed out one aspiring prospector who snuck in through the jungle forty-two times before Curio relented and issued the man a card. The Brazilian Senate president wrote a personal letter to Curio asking him to issue a card to a friend, and, as the story goes, Curio returned the letter to the sender with a simple answer: no. He set up a free health clinic and a malaria control post, a post office, telephone lines, and a government store where everything sold at cost. And he required that all gold be sold at the government office at market price.

Most significantly, he left the land in the hands of the miners. He and his staff of 150 soldiers, engineers, and geologists quickly demarcated ownership of individual plots of about sixteen feet by sixteen feet based on occupation at the time of Curio's arrival. No miner was allowed ownership of more than one plot; if they had more than one, they had to choose and turn the others over to Curio for redistribution through a lottery system. (Nonetheless, enterprising miners created minority interests in their plots and swapped them with others to spread their risk.) The system offered fairness, a concept that had eluded all these have-nots throughout their lives. It also offered certainty in ownership as well as an opportunity for a labor system. The owners needed workers to dig the plots, to carry away the dirt, and to sift through it. Curio capped the population at forty thousand.

Many miners wore T-shirts with Curio's likeness. Alongside the letters in his name was a motto, a slogan of manhood barked by many a high school football coach: C(oordination), U(nity), R(espect), I(dealism), O(rganization). "For us, God is in Heaven, and Dr. Curio is on earth," one miner said.

Curio, too, served as an important symbol in Brazil's transformation. Born in the interior state of Minas Gerais in 1934 as Sebastião Rodrigues de Moura, he earned his nickname as a tough boxer when he was young—a *curio* is a small and aggressive black bird. He graduated from the national military academy and was trained in intelligence and counterinsurgency work, as well as in jungle warfare. When leftist guerrillas first started appearing in the Araguaia Valley region of the Amazon in 1969, with an agenda of fomenting a Maoist/Castro agrarian revolt against the military government, Curio led the counterattack. The episode, cloaked in mystery, is a historical reminder of brutal military control. Several thousand troops spent three years hunting no more than a few hundred guerrillas, who never managed to incite rebellion among the local population. Reportedly most of them were killed after being tortured. As recently as 2004, the press reported the discovery of graves of headless victims.

Due to Curio's experience in the region, the government employed him as a troubleshooter to resolve multiple land and social problems on the frontier. A shadowy James Bond figure without an identifiable chain of command, he used the titles "Doctor" and "Major" interchangeably, often inventing pseudonyms and rarely wearing a uniform. A bishop in the Amazon once refused to keep an appointment with him, saying he didn't meet with birds. Curio sent word that he wasn't a bird but Major Marco Antonio Luchini, a name he borrowed from his wife's family.

Then, Serra Pelada changed him. In November 1980, President

João Figueiredo, the last general to run Brazil, came to Serra Pelada to witness the phenomenon of so many men living so close together in peace and industry. What he saw moved this taciturn general to tears. He promised that the mine would remain in the miners' hands.

That promise secured Curio's standing among the miners and caught the attention of Figueiredo's advisers, who were beginning to plan the transition from military to civilian government. Their goal was to keep power within the Social Democratic Party (PDS), the parliamentary arm of the military. The congressional and state elections in 1982 would be the first part of the transition. Who better to run for Congress from this area than Curio? In order to take advantage of Curio's popularity, the government lifted controls on the number of miners allowed in Serra Pelada. The population swelled to 116,000, and Curio won office easily.

Having achieved electoral victory with Curio, Figueiredo broke his promise to the miners. Under pressure from the Ministry of Mines and Energy to erase the precedence of individual ownership of natural resources, and under pressure to increase the yield through mechanization, the government in 1983 proposed taking over the mine and transferring the miners to other *garimpos* in the Amazon. Curio was caught between the miners and the military.

Looking back twenty years later, Curio is more bemused than troubled about that time. He had had to make a choice. "They asked me to burn down this town," he told us. He meant the eponymous town of Curionópolis, about twenty miles from Serra Pelada, where we were visiting him to see how age had treated a legend. He sat beneath a black-and-white photo of himself in which he was surrounded by scores of mud-caked men. A handsome female aide in her mid-twenties now sat at his side, occasionally holding his hand. Up close he showed what age does, even to the fiercest of men. His mane of unruly dark hair, now razor cut and dyed orange or blond, depending on the angle of light, framed his once angular but now puffy face. His eyes watered. His movements were elderly, slow.

Curionópolis had arisen spontaneously to serve the needs of the miners after the discovery of Serra Pelada. Wives and children and prostitutes found refuge there after Curio's arrival. It's where celebrations could take place with alcohol. It grew up on a promise-economy, where bills mounted until a miner could pay them off. When Curio received the orders to wipe it out, there were nearly one hundred thousand people in Curionópolis.

"Burn the town?" Curio recalled. At the time, the order hadn't been so preposterous. He had been a soldier, and those had been his orders. But he had also become a politician, and these people had been his constituents. He explained, "I came here to see what they wanted me to do, and children came out into the street. 'Curio! Curio!' they shouted. They followed me everywhere." He paused. "I couldn't do it. I couldn't burn the town. Instead, I built a road. I built a school. I tried to do for them what I had done for their fathers and their husbands."

Curio's next move showed how quickly Brazil was changing from an autocracy to a democracy, which Serra Pelada, in no small part, inspired. Curio introduced a bill in Congress to keep Serra Pelada in the hands of the miners, he organized a regional branch of the national miners' union, and he sought a coalition with the powerful metalworker unions of São Paulo. Interestingly, the union leadership included José Genoíno, an Araguaia Valley guerrilla leader Curio had captured ten years earlier. The politician organized the miners to travel to Brasília en masse and lobby for his legislation. He challenged the takeover in the courts and planted stories with the press. Serra Pelada became the first national cause to utilize all the nascent institutions of the democracy soon to emerge.

The legislation passed, though President Figueiredo vetoed the bill. But Figueiredo then decided to postpone the actual takeover. He wanted the civilian government, expected to take power in 1985, to make the final decision. But the miners were not satisfied, and in

1984, a rebellion erupted in the area. Miners staged mass demonstrations, demanding long-term ownership. They blockaded roads, cut telephone lines, and held federal workers at Serra Pelada hostage. Curio intervened and persuaded the miners that military intervention, followed by bloodshed, was imminent; he made the same plea to the government. Figueiredo ultimately relented and signed a bill giving the miners ownership for the next five years.

Curio left Congress in 1986, settled in Curionópolis, and continued to work to establish the miners' control over Serra Pelada. In 1987, the issue of control resulted in a violent confrontation. The miners seized a railroad bridge near Maraba, and the state governor ordered the military police to evict them. According to Amnesty International, as many as eighty-six miners were shot and thrown from the bridge. The miners seemed to be in a perennial state of conflict over ownership issues. In October 1996, the federal government sent in one thousand troops and sixty-three federal police to quell the Serra Pelada Liberation Movement, formed by miners to impede the exploration of the area by the government-owned mining company. All the while, Curio oversaw the miners' various lawsuits. One suit sought fifty million dollars in damages from the government for understating the purity of the gold bought from miners. Another suit sought to establish landownership, a campaign that ended successfully in 2002 when Congress restored the rights to a cooperative of miners Curio had formed.

Curio returned to politics in 2000 when he was elected mayor of Curionópolis, although he continued to focus on miners' rights. When we visited, he was negotiating the sale of the mine to a Michigan-based company, Phoenix Gems Ltd., which reportedly would pay the miners $240 million in cash, plus future royalties. Curio estimates that 500 tons of gold remain in the ground. By comparison, from 1848 to 1856, the California gold rush averaged 80 tons of gold annually, and the Klondike rush 42 tons per year in its peak years. Although in its peak year of 1983, Serra Pelada yielded 14 tons of gold, the output of the entire country reached 120 tons in 1987, which placed Brazil behind only South Africa and the Soviet Union in gold production.

We asked Curio what had happened to the men we had met during our first visit—the ones who bought Rolexes when they couldn't tell time; those who bought cattle ranches, unable to understand the paperwork associated with business; and those whose idea of a one-night celebration ended up in serial marriages and expensive divorces.

He laughed at the memories of that time. "No one has his money still. Almost no one. It was like a fever. It made them all sick. But at least they had their moment. I think the ones who made out the best were the guys who hauled the dirt. They got paid ten times what they would have elsewhere in Brazil. Enough to live. Not enough to be rich."

If the sale of the mine to Phoenix Gems goes through, some of these men stand to be rich again. Many miners, in need of instant cash and skeptical of the lawsuit, sold their shares to others, including Curio. "I don't have as many shares as some," he explained. "But I don't have as many years to spend the money. And, also, more important, I am miner number one on the list of shareholders because that's what the others wanted."

During our visit, we attended a soccer match with Curio. Fittingly, the game was between two teams of adults obviously past their prime. Cindy Lauper's "Girls Just Want to Have Fun" blared from speakers, and crowds cheered in an atmosphere reminiscent of a country fair.

Curio seemed to appreciate our company. Perhaps we reminded him that he is more than just a municipal mayor in a sprawling Amazon town. As we sat with him, well-wishers visited him on the enclosed platform at midfield that served as the VIP box. They asked for his help in cleaning up garbage on their streets, or they asked whether a particular road would be paved soon.

For us, Curio had been a generational figure, someone who embodied a particular moment in history—the earliest flicker of national self-discovery—like a Joe Louis or a Eugene McCarthy, whose personal impact could never be fairly conveyed by history books. Once, he hunted Communist guerrillas in the rain forest. Now Communism has almost vanished, the rain forest has given way to highways, and the mine is being sold to an American company. Everyone carries a cell phone. Blooming metallic sunflowers line the tops of even the most basic wooden houses. Broadband is coming to Curionópolis. Just another reminder of how fast this world is changing.

CHAPTER 9

OIL: SPOILER OR SAVIOR?

—

In 1980, we first saw Serra Pelada as a pink dot on satellite maps used to track deforestation. Solid green indicated intact forest, and pink showed transitory invasions. Black meant permanent settlement. The color patterns varied from map to map, depending on which satellite took the pictures and which office was reviewing them. At that time, though, every map was dominated by green in the western part of the Amazon.

Carlos Marx was then in charge of Brazil's fledgling satellite mapping program. The international environmental movement had just begun to appreciate deforestation of the Amazon as an easily understood indicator of the pace of nature's demise. Statistical evidence showing accelerated deforestation had never been available. Technology now could be used to judge Brazil's guardianship of this resource, and it could provide empirical data in order to garner financial support for preservation efforts. Over time, the annual release of this data would become tantamount to environmentalism's annual physical, the results broadcast and reproduced around the world.

At the time, Marx was intrigued by the gadgetry of the process, not the politics. Deforestation amounted to no more than about 3 percent of the Brazilian Amazon. Although some scientists were alarmed at exponential increases due to the pace of road building, there was so much green on these maps that it was hard to contemplate a significant shrinkage. "The majority of the rain forest should last forever," Marx told us. He pointed to the heart of the state of Amazonas, the center of the green. "No one can get there," he assured us. While he conceded that deforestation would eat away at the edges of the area, he was confident that the jungle's inaccessibility would be its salvation.

In 1980, we met only one person who challenged Marx's hypothesis. Dorival Knipboff, a millionaire builder from the far south of Brazil, told us he planned to clear about 1.5 million hectares in the heart of the rain forest to create a palm oil plantation. With little refinement, his oil could be used as a substitute for diesel. The wood could be transformed in a giant alcohol refinery. "We have the technology. We will just take our time. We expect to be there for many, many years," he boasted.

Twenty-five years later no one had heard of Knipboff. No one we spoke to in São Paulo, where we first heard his name at the trade association of Amazon businessmen; no one in Brasília; and no one in Manaus, the nearest city to his dream project. Most likely, he never cut a tree. Yet when we checked recent satellite maps, we saw pink precisely where Knipboff had planned to build his plantation, the very spot indicated by Carlos Marx when he said, "No one can get there."

We went to find out what was there.

The flame on the horizon is startling, a tight orange cone shimmering over the tree line. After flying for almost two hours southwest from Manaus with nothing but trees and an occasional snaking brown river underneath, any sign of civilization is satisfying. The fire's source becomes clear as we approach: a sprawling series of white chimneys, part of a high-tech industrial complex that looks like a secret military installation. An army of workers in orange jumpsuits moves through a maze of pipes and steel towers and low squat buildings. We hadn't seen a town for hundreds of miles in any direction, not even a road, except for the spine of black pavement we spotted as we approached this clearing. A wildcatter from Oklahoma exploring for oil in the Peruvian Amazon once said to us, "As a general rule, you have to remember the good Lord was a fine man, but he picked some godawful places to put oil." This place was one of them.

This oil and gas field at the headwaters of the Urucu River lies almost dead center in the South American continent, surrounded by primary rain forest for hundreds of miles in all directions. If there were a part of the Amazon that even the most worrisome environmentalist considered impenetrable, this would be it. It's the only pink splotch on that spillage of green, but it means that no place is out of reach any longer.

The factory's flaming towers are part of a typical oil and gas plant, complex in design but simple in purpose: to sort the hydrocarbons pulled from wells over the 2,500-square-mile field and route them into a pipeline. It's the standard petrochemical extraction process, only exceptional because no one thought it would ever happen in the center of the Amazon basin.

Oil, like gold in the Amazon, is a commodity many have thought abundant, though eternally elusive. For almost a century, Brazilian geologists assumed it had to be there: primary rain forest sitting in a bowl-shaped basin was an ideal landmass for petroleum. The center of South America is believed to have been an ancient plain, then an inland sea. The Amazon River once flowed to the west, draining into the Pacific Ocean, until the Andes Mountains rose up and tilted the tectonic plates the other way. In the 1960s, oil was discovered in the Ecuadoran Amazon in what would have been the delta of the westward-flowing river, or perhaps the shoreline. After much poking into the Peuvrian forest floor, significant amounts of oil and huge amounts of natural gas were discovered. But until recently, Brazil had come up dry.

Oil exploration in the Brazilian Amazon began in 1917, with a handful of true believers traversing the basin hectare by hectare. Oil prospectors traveled weeks by boat or canoe, and hiked into the forest to punch holes and take soundings. On paper, the Solimões basin, south of the main river, was an altogether promising place to look. Framed by underground structures known as the Iquitos Arch to the west and the Purus Arch to the east, it seemed to sit on a rich lode of Devonian shale and Carboniferous sands. But again and again, the efforts produced nothing. It was like sticking pins into a continent-size carpet. By the 1960s, Walter Lake, a legendary South American oilman, reportedly vowed, "I'll drink any oil you find in the Amazon."

Petrobras, the Brazilian national oil company, found the first hint of gas in 1978 near Urucu. The magnitude of the find was not clear until 1986, and it took another two years to start production. It's estimated that there are at least one hundred billion cubic meters of gas and eighteen million barrels of oil in the Urucu region.

"This is not Saudi Arabia, but for Brazil it will be very helpful," Ronaldo Coelho, the manager of the site, told us during a visit. The Amazon gas reserves are the second largest in the country, and they will play a significant role in Brazil's effort to become self-sufficient. By 2007, Petrobras expects to supply all of Brazil's petroleum needs through domestic sources. This will be possible not only because of the development of oil reserves but also, and more important, because there is a decreased demand for oil due to natural gas development and alternative fuel sources. All gasoline sold in Brazil already contains 26 percent ethanol, primarily made from sugarcane. In 2004, Brazilian farmers produced 385 million tons of sugarcane, from which refiners made four billion gallons of alcohol fuel—enough to replace 460 million barrels of oil. Ethanol also has become a boon for Brazil's balance of payments account, as exports are expected to rise from $600 million in 2005 to $1.3 billion in 2010. This rise will primarily result from Sweden's and Japan's commitments to the fuel to help them meet their emissions quotas under the Kyoto Protocol.

Energy independence for Brazil has long been a goal. The oil shocks of the 1970s, first in 1973 after the Mideast War and then following the Iranian revolution in 1979, single-handedly snuffed out the Brazilian Miracle, the economic Camelot that Brazil enjoyed in the early days of the military government. The effects of that disaster, seen in the crippling foreign debt (40 percent of the country's foreign exchange income was used to import oil), still plague the economy. The oil and gas reserves in the Amazon, therefore, are of critical importance for the area's growth. Their proximity to the industrial zone of Manaus, which provides significant export revenues and employment, is an asset in attracting capital investments. Even though the exploration costs have been high—from 1954 to 2004, Petrobras spent $7.5 billion—the geology of the area minimizes operating costs. The hydrocarbons are high quality and easily recoverable. The fields are at a relatively shallow eighty-two hundred feet and are under enough pressure that the oil and gas surge up easily through the pipes. The crude is unusually pure, bubbling out of the wellhead like espresso. "You could practically strain this through your handkerchief and put it in your gas tank," said Coelho as he rubbed some between his fingers. "The only issue is how to get it out of this site to a market. And that's a political problem, not a technical one."

It's a big political problem indeed. In 1998, Petrobras completed a set of pipelines of about 175 miles from the fields to the Amazon River port of Coari, where the oil is loaded onto tankers for the sixteen-hour journey to Manaus. A second pipeline moves liquefied petroleum gas, known as cooking gas, which also is sent to Manaus by boat. This gas ends up in the blue containers of Fogas, a company owned by the Benchimol family, and provides the main source of cooking fuel throughout the region.

The daily output of the wells is sixty thousand barrels of oil and 353 million cubic feet of natural gas. In January 2005, after five years of public hearings and lawsuits, Petrobras began construction on a 250-mile extension of the gas pipeline from Coari to Manaus, which will allow natural gas to travel from the field to the city. In order to secure permits for the new pipeline, Petrobras agreed to reroute the lines in order to minimize environmental impact. The company also agreed to provide gas pipelines to small regional towns, to hire local workers, and to provide the towns along the way with Internet and phone service. The give-and-take of this process, which included environmental impact statements, provided another contrast with the large-scale projects of twenty-five years earlier, which had no grassroots involvement, only military decree.

Now attention is focused on a second pipeline route scheduled to commence in 2007. This route will travel 320 miles from Urucu south to Porto Velho, the capital of the burgeoning state of Rondônia. Whenever an access route has been created in Rondônia, a spontaneous influx of immigrants hungry for land has emerged, and there has been little state control over development. The same warnings over the Manaus pipeline are echoing. Environmentalists see it as a deathblow to the remote western jungle, fearing that this pipeline will open a seam of entry to empty forest and protected Indian lands, clearing the way for a torrent of loggers, miners, and cattle ranchers. The pipeline controversy has complicated the traditional debate about the projects that deforest and the scant benefits they provide. The construction of these pipelines will alter the rain forest but will also generate energy for millions of people hundreds of miles away. Nearly two million people live in Manaus alone, and they need energy. Blackouts rotate through the city daily. Lack of energy has retarded factory construction, holding back employment expansion. When Brazilian president Lula da Silva approved the pipeline to Manaus in the spring of 2004, he said, "If people want development that preserves the environment, we have to have energy. It's no good people saying the Amazon has to be the sanctuary of humanity and forget there are twenty million people living there."

The sign over the Urucu command post reads PROGRESS IN HARMONY WITH NATURE. The fields were developed with what Sven Wolff, the director of the project, calls an "offshore model"—as if they were in the ocean and not on land. "We want to make the smallest footprint we can." There are no roads into Urucu. Heavy equipment must be barged in. The eighteen hundred workers come and go on three-a-day flights for shifts of two weeks on, two weeks off. They live in a prefab compound resembling a military base, replete with a health clinic and a library with Internet connections. The wastewater pumped from the fields is treated. Trash is recycled or shipped back to Manaus. There is even a sprawling plant nursery that has already been the source of a million seedlings replanted wherever the earth has been torn up. Even the hundreds of well sites cause minimal impact—twenty-by-twenty-foot concrete slabs with a few pipes protruding from the center. It's easy to envision Petrobras capping the wells and having the forest reclaim the site in a few years.

According to Wolff, the pipeline to Manaus will be built with the same principles, slipped through a small slit in the forest with no access road alongside. "We'll have helicopter pads along the way for maintenance, but this will not become a highway for colonists," Wolff said.

Petrobras, by piercing the heart of Amazonia and profiting there, has shown that it is not such an inaccessible place. Or such a forbidding one. "We have very little injury or disease," said Ronaldo Coelho, the manager of a project that some of his friends think is as remote as the moon. "We haven't had a single case of malaria or any of the exotic diseases people worry about. Maybe one or two snakebites over the years. For people in the oil business, or maybe any business, this is not a very difficult place to work."

The opposition to the Urucu pipelines is not just about whether they can be built with environmental safeguards. The real fear isn't about the course of the pipe but about the product it carries. Oil brings modern civilization. Air-conditioning. Cheap gas. Industrial plants. Thriving cities. Brazilians, including those in the Amazon, strive for those resources. Are those benefits more important than preserving the hunting grounds of a five-hundred-member Indian tribe? Or a tract of forest that may or may not harbor the cure for cancer? These questions will determine the future of the Amazon. Cost-benefit analyses now can be made, whereas twenty-five years ago almost no one could point to reasonable trade-offs between development and deforestation.

CHAPTER 10

WHERE DREAMS NO LONGER DIE

—

In the early 1980s, the town of Sinop in the state of Mato Grosso, 193 miles north of the capital city of Cuiabá, was selected as ground zero for a revolution intended to make oil irrelevant. Some overly imaginative Brazilian scientist had postulated that manioc, a tuber like a potato, could replace oil as an all-purpose energy source. Charles Wagley, whose 1953 *Amazon Town* provided the first reliable descriptions of Amazonian daily life, noted the ubiquity of the plant and its importance as a staple in the region. "Manioc is a hearty plant," he wrote, "well adapted to the tropics and to the leached tropical soils. It grows in a variety of soils. It resists insects, especially the sauva ant, better than most crops. It prospers either in heavy or in light rainfall."

Through modern alchemy, this subsistence crop would be transformed into a gasoline substitute. Oil? Who needs it? Grow manioc! After the Mideast oil price hikes in the mid-seventies drove daggers into the Brazilian economy, this mania spread. The government was desperate to find alternative fuel sources. Sugarcane, the base of ethanol, was already showing promise. There was a predisposition for cockamamie schemes.

A lot of people fell for it. Sinop attracted thousands of farmers in search of opportunity, get-rich hucksters hoping to profit from these strangers in a strange land, and small-town merchants who set up shop to service this economy. Officially founded in 1972, Sinop was a mammoth science-class experiment open to the public. By 1980, it had attracted twenty-five thousand inhabitants, most of them dedicated to a single purpose: turning the Amazon soil into black gold. Loggers razed the forest, sawmill owners sold the timber, farmers planted the land, laborers harvested the manioc, and factory workers turned it into a miracle potion.

It was a quintessentially Brazilian adventure—a revolutionary idea brought to life on a grand scale in an exotic place. Manufacturing ethanol from manioc combined twenty-first-century vision and science with Brazil's abundant land. It may have been yet another instance of public money wasted, but underneath the folly lay a serious effort to create a uniquely Brazilian success story. The Amazon landscape provided an endless source of raw materials for a finished product that would have endless demand. If it worked, this sea of cheap land would become a sea of riches. Brazil's future appeared in the thirty-three-million-dollar government-subsidized alcohol distillery that loomed, like a large Erector set, over tens of thousands of hectares of planted manioc in a once-upon-a-time rain forest.

Sinop also represented a new milieu of opportunity. This frontier project was different from a gold strike, which attracted monomaniacal antisocial miners, or an oil find, which led to government or multinational takeover and control. When the Brazilian economy floundered in the late 1970s, thousands of farmers in the south went to the Amazon in search of cheap land, confident that they could make it arable despite their having no experience with this soil nor models to copy. The real estate promoter behind Sinop advertised the project as worthy of these farmers' skills, accommodating of their ambition, stable and secure for their families, and steeped in the ethos of a meritocracy.

If the Amazon were ever to be settled in a sustainable way, it would have to happen in communities under some recognizable and approachable authority—and it had to start with certainty in land title, an elusive concept on the frontier. Sinop was a private colonization project owned by a single individual, Enio Pepino, who might have been a great snake-oil salesman—selling a manioc-to-ethanol factory, like a golf course, as the catalyst for real estate development. Nonetheless, a private governing authority, selling off clearly demarcated plots of land, enhanced Sinop's chances of success as a permanent community. The town would not be a happenstance migration stopover following the construction of a road. Government agencies would not race to keep up with land titles, enforcement of environmental regulations, or the establishment of social services. The profit motive behind Sinop, selling land to settlers who then had a vested interest in the enhanced value of that land, gave the settlers incentive to stay and prosper rather than slash and burn the land and then move on.

Sinop also straddled the planned road from Cuiabá to Santarém, BR-163, which at the time was yet another work-in-progress artery planned to connect the south of Brazil to the Amazon River. Sinop's economic promise meant that people might stop on the road and plant roots, something that hadn't happened in the decade since the east-to-west TransAmazon Highway had been built. That road had simply created a vein into the jungle by which unskilled farmers traveled from place to place, leaving deforestation and degraded land in their wake. The dream behind Sinop was an industry capable of supporting a second generation of settlers. These farmers were expected to create institutions such as schools and health centers and other municipal services.

In the Amazon, roads of ambition wash out as regularly as roads of red clay, so we weren't surprised on our return visit to learn that Sinop's manioc experiment had failed, just another statistic in the Amazon's triumph over attempts to mass-produce a single crop. Manioc may have powered a car somewhere, at some time, but only as a curiosity and not as a challenge to petroleum's hegemony. Pepino's factory of rusty pipes with paint-chipped vats and control panels enveloped by cobwebs dominated the Sinop skyline like a silent prehistoric dinosaur. Our taxi driver didn't even know what it was. Something about this failure differed from the others we had witnessed throughout the Amazon. Few of Sinop's citizens with whom we talked twenty years later knew the distillery had even existed, let alone understood why it had been built in the first place. Sinop now has 120,000 citizens living along paved streets in neat subdivisions, waiting patiently for the completion of the new airport terminal and the new forty-six-store indoor shopping mall. The dream of manioc had died, but the city had persevered.

Maybe the string of failures that long littered the Amazon is coming to an end. We had spent months visiting the notable (and tragic) follies during research for our first book, which once had a working title of *Where Dreams Died: Man's Fate in the Amazon*. In Porto Velho, the capital of the southwestern state of Rondônia, we'd visited the planned terminus of the Madeira-Mamoré Railroad, once connecting the rubber tree production of landlocked Bolivia to a port on the Madeira River, which in turn connected to the Amazon. Brazil and Bolivia agreed to build the railroad in 1867 as part of a friendship treaty, but it wasn't finished until 1912. Ironically, this was the year the price of South American rubber peaked. The price steadily declined, and the railroad passed into obscurity.

There are few places on earth where so many died without war for such futile reasons. "They say there is a skull resting on every tie," said our guide, Silas Shockness, whose father had come from Grenada in 1914 to work on the railroad. With 1,630 ties for each of the 227 miles, that would be almost six hundred thousand skulls (the true figure is probably a fraction of that). As most of the workers were poor immigrants from countries ranging from China to Greece, there are no reliable records of the number of victims, most of whom died from malaria and yellow fever.

The Opera House in Manaus stands as perhaps the Amazon's best-known monument to grand ideas gone awry. "Before rubber, Manaus was just a tawdry jungle town of a few thousand souls," wrote Jonathan Kandell in *Passage Through El Dorado*. "But by the turn of this century, its population had swelled to seventy-five thousand and, for a city that size, it flaunted more luxury than Paris." In the late nineteenth century, when rubber became what manioc was supposed to have been in the late twentieth century, the people of Manaus could afford anything. They built an ornate opera house, which defined that moment in history: the chandeliers came from France, the marble from Italy, the wrought iron staircases and balconies from Britain. The seats, made from local mahogany, were carved and finished in Europe, and covered in velvet. The green and gold cupola dome contained tens of thousands of tiny tiles imported from Europe. The citizens of Manaus brought Anna Pavlova to dance and Jenny Lind to sing for them. Completed in 1896 after fourteen years of construction, the opera house has long been the best-known symbol of occupation of the Amazon, serving as an object of derision rather than accomplishment. Its declining state throughout most of the twentieth century reinforced Betty Meggers's warning that modernity and the rain forest could never coexist. The majestic building, which had once gleamed like a holy mosque, had a sadness to it, like beauty's death from disuse. Recently, it has been renovated, an appropriate symbol of Manaus's resurgence after ninety years.

Another noteworthy skeleton of grandeur lies about 300 miles away on the banks of the Tapajos River, about 125 miles south of that river's intersection with the Amazon at the town of Santarém: the site of Henry Ford's folly, Fordlandia. Ford came to the Amazon in 1927 desperately looking to escape the clutches of the British-Dutch rubber cartel that was manipulating the supply and price of Asian rubber. The Europeans came to dominate the rubber market because of Henry Wickham, an Englishman who single-handedly impoverished the Amazon through an act of bio-piracy. In 1876 he stole rubber seeds and took them to Malaya, where they thrived in the tropical climate and soil, free from the natural pests that had made it impossible to grow the trees in plantation form in the Amazon.

Brazilians hoped that Ford's ingenuity would help them grow trees in volume and recapture control of the world market. Ford readily responded, as he needed a cheap supply source. Brazil gave Ford nearly 2.5 million acres of land and most of the profits he could make from it. But the Amazon was not an environment where trees could be produced on a Ford assembly line; they were killed in short order by a fungus that spread from tree to tree and by uncontrollable soil erosion.

In 1938, Ford swapped approximately seven hundred thousand acres for another site closer to Santarém, called Belterra. He sent researchers to Asia for disease-resistant strains of trees that were then grafted onto Amazon varieties by an expensive and ultimately fruitless procedure. Ford tried chemical sprays against the pests, but those failed, and though he managed to plant over 3.6 million trees, he lost almost all of them to nature. He declared defeat in 1946, having lost nearly thirty million dollars in the process.

The most intriguing failure in the history of the Amazon, however, may belong to Daniel Keith Ludwig, because of the size of the project, the scope of Ludwig's vision, the amount of money involved, and his penchant for secrecy. Ludwig, a self-made billionaire from Michigan and reportedly once the world's richest man, made his fortune by inventing the supertanker and then assembling a fleet of them. He had no publicly recognizable corporate entity, and he had a publicity phobia, both of which added to his mystique. In the mid-1960s, Ludwig became convinced that a shortage of wood would grip the world before the end of the twentieth century. Increases in population would cause demand for forest products to rise, and pressure for land for commercial and residential use meant supply would dwindle. A new source would need to be found.

Ludwig's solution for the shortage was to find a fast-growing tree suitable for plantations located in underdeveloped areas where land was plentiful and cheap. This tree would also have to readily yield the lumber and pulp the world would soon so badly need. The billionaire hired scouts to roam the globe searching for a tree that could grow quickly in a tropical climate. It took ten years, but he found the melina tree, which, his agronomists assured him, grew a foot a month. The tree, native to Southeast Asia, had been transplanted to Africa by the British, who used it for fuel and mine-shaft supports.

Ludwig's search for land came to the attention of Brazil's ambassador to the United States, who lobbied Ludwig that Brazil could meet the four prerequisites that he had set for the project: large uninterrupted expanses of land, a cheap labor source, proximity to a deep-water port, and a stable government. That the land was in the Amazon, which had never succumbed to the hand of man, meant little to Ludwig. He had never failed at anything before.

The Jari Project (named after the river flowing through it) started in 1967 and occupied land larger than Connecticut and Rhode Island combined. Ludwig built more than three thousand miles of road, thirty-seven miles of railway, and a deep-water port. He planted 260,000 acres of trees. In anticipation of the harvests, he constructed two sixty-six-million-pound seventeen-story factories, one a pulp mill and the other a power plant, for $269 million. Each was shipped, fully assembled, to Jari from Kure, Japan, a three-month journey of more than fifteen thousand miles. The press ran photographs and stories about the buildings on the barges, a modern city cast adrift. After being maneuvered into a holding lagoon whose entrance was shut off and whose water level was raised, the factories were then floated onto thirty-seven hundred pilings shaped like cigar butts and cut from rot-resistant massaranduba wood. The factories then sank into place as the dike was opened and the water returned to its original level. The placement was so accurate that joint connections between the plant and the in-place machinery at the site were never off by more than 0.37 inch. Ludwig also had his share of luck. He found the world's third-largest kaolin mine on his land, and he built a $25 million processing plant to prepare it for export.

But the Amazon also thwarted Ludwig. His bulldozers damaged the delicate soil when they cleared it, so all clearing had to be done carefully and expensively using chain saws. The melina tree didn't adapt to the sandy soils of Jari, so Ludwig switched to a Caribbean pine tree, which didn't grow nearly as fast. He also envisioned the Amazon as a potential breadbasket, but his vast rice plantations could only be seeded, fertilized, and sprayed by air, which became too expensive. And it turned out the soil couldn't sustain multiple plantings of rice. Ludwig fired thirty directors in thirteen years, believing the problem to be managerial, not environmental.

Most significant, he fell prey to the opening up of the press in Brazil, which began to question why an American owned a country-size piece of land within Brazil, why thirty-five thousand people worked for him, why he had twelve airplanes, and why he was hostile to media and government visits to his property. With costs rising and without any support from the government, Ludwig essentially dismantled his empire to support this project. He sold off his worldwide collection of hotels, mines, orange plantations, and oil and gas properties, but all that money couldn't rescue the endeavor. He finally sold his interests to a consortium of Brazilian enterprises in 1982 for less than a third of his investment, a loss of nearly a billion dollars. He died in 1992 at the age of ninety-five.

Sinop, on the other hand, was a failed experiment that somehow managed to succeed. While founded on the same type of hubristic vision as Fordlandia or Jari, Sinop differed in one critical regard: on-site individual ownership of clearly titled land.

It's easy to tell from the heavy truck traffic along the encrusted red clay BR-163 highway bisecting Sinop that prosperity won't be transient. Soon all of BR-163 will be paved, connecting Cuiabá to Santarém and providing an outlet for the area's agricultural production. Land values along the entire road have soared, and that trend will continue. But land speculation alone does not explain why Sinop has prospered.

Jaime Luiz Demarchi, forty-two, owns several farms and runs a machinery shop on the service road along BR-163 just outside of town;

commerce there is dominated by rows of repair shops. "I arrived here on June 7, 1974." (Most successful immigrants we met throughout the Amazon knew the exact day their families arrived.) "We came from Paraná at the very beginning. There were only thirty-eight houses here. We had nothing. To survive we had to hunt and fish in the jungle." They sold their home and all of their belongings in the southern state of Paraná and bought a plot from the colonization company. They suffered through years of subsistence living. Yet they remained confident they could figure out how to work any land, be it in the jungle or elsewhere, as long as they had the time to learn from their failures and the opportunity to share information with their neighbors, also struggling on unfamiliar soil.

"We really had no choice but to succeed," said Jaime. "No one I know went back to the south. To what? The only people who have left Sinop since we arrived went north to open new frontiers. My family and the families who came here had a philosophy that the more difficulty, the better. They have a pioneering spirit and courage, and they are smart. They may not be smart in school but they are smart in how to survive."

Demarchi laughed at the memory of the mania of manioc. "It was a great idea, but it made no sense. They built a factory and then expected that because there was a factory they could grow manioc. That is crazy."

It took years for Jaime and his neighbors to understand the soil: what crops worked, what rotation they needed, what fertilizer worked best. "We now have rotations that go through soy, cotton, wheat, and rice," he said. "And each of those plantings may need a different seed, depending on where they are, what the soil is like, what the rotation is going to be. We can figure these things out because we have the most modern technology in the world. Right here in Amazonia."

In his machine shop, Demarchi has a computer with broadband access, which allows him to share information about seed varieties with research organizations, to download weather information, to buy and sell equipment, and to keep up with the commodities markets. He has a cell phone. The advantage that American and European farmers once had over Demarchi—access to information and technology—is gone. The competitive matrix now tilts in his favor because he also has a year-round warm climate, abundant rainfall, and plenty of land.

For the Demarchi family, the push toward the frontier stopped in Sinop. This migration was a sea change from the belief that Amazonia was only good for slash and burn agriculture. With three children—ages three, twelve, and sixteen—Demarchi has found a home where he intends to stay, vowing to resist selling his land along BR-163, which doubles in value every two years.

Angelo Carlos Maronezzi, forty-four, another early immigrant, came to Sinop in 1974 with a suitcase of clothes and "the promise of a better life." His father, also a tenant from the south, used his savings to buy land in the early development of Sinop and "put our whole family to work with the soil."

Maronezzi now owns a large ranch about ten miles outside of Sinop, where he experiments with varieties of rice and soy, trying to match the right crop with the right soil. Although soy brings more income, Maronezzi said, "we have learned our lesson here not to have a one-crop economy, which will make us vulnerable to losing it all."

The difficulties of learning the ways of tropical agriculture had humbled Maronezzi and his colleagues. "Twenty years ago, farmers did not have any environmental conscience," he said, appreciating neither the difficulty of succeeding in the Amazon nor the consequences of their efforts. "They would clear right up to the river and dump poison into the water, and it was not only bad for the environment, but it ruined the possibility of using the land for a long period of time. Farmers are trying to stop this deforestation where it has no purpose, because it is not helpful to our business. I am trying to determine how to grow the maximum harvests with the least amount of deforestation."

The challenge for Sinop and the other colonization towns like it will be to make success endure. Maronezzi pointed out that "anyone can plant one or two years"; these farmers have an expectation of multiple generations using this land. He conceded that trial and error, the only way to proceed, constantly yields different results, as what works for fifteen years might not work after fifty. The collaboration among the farmers and their ability to call on resources throughout the world has changed the dynamic toward creativity and adaptability. Time will tell whether that will be enough to deal with the defenses of an endlessly creative and adaptable natural environment.

CHAPTER 11

FROM POVERTY TO SOVEREIGNTY

—

The BR-163 highway runs like a timeline through the past thirty years of Amazon history. In 1973, when the road was finished, the only ostensible planning that had gone into it was where it would start and where it would end: Cuiabá and Santarém. Another vein facilitating the flow of immigrants had been opened, satisfying the military government's objectives. Where these people settled, on whose land they settled, what they did with the land they occupied—these issues mattered little to the generals. Even a modicum of planning could have prevented the chaos that ensued along the road, and spread from it—land wars, destructive cattle ranching, and mahogany poaching, for example. BR-163 served its geopolitical purpose for the government in Brasília by flooding the area with immigrants. It did little for those on the ground. Sinop and a few other private colonization projects were exceptions.

The Lula government, which took office in 2003, seized the opportunity of the imminent paving of BR-163 to try to undo some of its original sins, or at least to avoid repeating them. The newly appointed Minister of the Environment, Marina Silva, put BR-163 on her priority agenda, recognizing that paving would attract a new wave of settlement. She perceived that the improved road also would facilitate a west to east development pattern. This pattern, she foresaw, would compound the deforestation resulting from the Belém-Brasília Highway and portend a barren corridor in the heart of Amazonia. In addition, the road would inevitably launch east to west deforestation, which would jeopardize areas that were still intact only because of their inaccessibility.

Silva advocated interministerial oversight for the paving project because she figured that competing interests would lead to compromises and would prevent a chaotic settlement after the initial construction. She brought the agricultural, energy, transportation, law enforcement, colonization, and environmental bureaucracies to the same bargaining table. She also made clear to the state governments of Mato Grosso and Pará that she wouldn't yield jurisdiction in the area just because they might succeed in raising private funds to pay for the paving of the road. She mobilized Brazilian and international NGOs to play a role in preparing the population along the road for the increased traffic and consequent development.

With the paving of BR-163, Silva hoped to revolutionize the planning process for national highways by emulating the model used for road building in her home state of Acre. The difference between the Acre model and the planning behind the major Amazon roads (BR-364, Belém-Brasília, and the unpaved BR-163) was that the military government, in a hurry to establish presence, built roads and then dealt with the consequences. The Acre model advocated planning for the arrival of the road before it was built. The latter meant setting aside forest reserves along the road to keep them off limits to arriving settlers; regularizing land titles in the vicinity of the road, where the density of population would be greatest and conflict most likely; educating the population in place to prepare them for the increased traffic, an influx of settlers, and a strain on existing resources; providing infrastructure for social services so that settlers would be inclined to remain in place and build, rather than slash and burn and move on; conducting soil research to provide advice for aspiring farmers; establishing monitoring systems to enforce restrictions on forest clearing; and allowing the Ministry of Justice and the state governments to plan for maintaining law and order.

The experience of the past thirty years showed that much of this planning would prove futile, but this didn't deter Silva. The expansion of the agricultural frontier in the region had made BR-163 a vital lifeline from the fields to the ports of Itaituba and Santarém, each of which had been transformed to handle bountiful harvests. Because of the road's economic importance, Silva had the upper hand: her environmental concerns would have to be heard, or the ranchers and farmers would not get their paved road.

Nor would they be able to circumvent her easily. Lula credits Silva, a comrade in arms of Chico Mendes and several other activists in the state of Acre, including Governor Jorge Viana, with the birth of the rural wing of his Workers' Party. She enjoys a strong personal relationship with her president, which arises from these common struggles. As important, Silva has become a rock star in the international environmental movement; her appointment comforted NGOs and other countries concerned with the Amazon's fate. Consequently, the government could not easily weather her resignation in protest about paving BR-163, and any agencies that might oppose her will be wary of domestic and international condemnation. Silva's strength is perhaps not in what she can do, but in what she will do if she doesn't get her way.

It was easy to appreciate Brazil's changes over the past generation when it came time for us to again visit "the most important person in the Amazon" in Brasília. During our first visits, that person had been Mário David Andreazza, a former colonel, whom we compared to "the highest-class Chicago ward boss, though with his tanned face, his wavy silver hair and fine-cut silk suit, he looked like a million bucks on Miami Beach." As Minister of Transportation in the early 1970s, Andreazza had overseen the construction of the TransAmazon Highway. At the time of our first visits, he had been Minister of the Interior, a position that gave him unchecked power over the Amazon. The states themselves hadn't received the autonomy that the 1988 constitution gave them (Rondônia and Amapá hadn't even become states), and Andreazza had complete control. We came to his attention at a time when, as we wrote then, he "publicly reversed his development-at-any-cost philosophy and claimed he was trying to inject rationality into the push westward." To prove the point to us, he made available planes, helicopters, and jeeps. Told us to go wherever we wanted. See whatever we wanted. Write whatever we wanted. No strings attached. We were in the right place at the right time.

When we returned to Brasília twenty-two years later to see "the most important person in the Amazon," we found ourselves with a frail brown-skinned woman who'd been changing bedsheets as a chambermaid when Andreazza presided. Born in 1958 in the state of Acre, Marina Silva was one of eleven children, three of whom died in infancy. Her father, a rubber tapper, left home for days at a time to ply his trade, an especially rough life of loneliness, poverty, and sickness. Silva grew up in a wood-plank house built on stilts under a palm-leaf-thatched roof—typical rain forest housing—without electricity, phones, or access to health care. As her father needed all of the family's labor to support his tapping activities, she never attended school.

In the mid-1970s, the government land agency divided up the land where Silva's family tapped trees, and gave small plots to the tappers and the thousands of immigrants then arriving from the south. This change in land tenure brought about a change in land use. Silva's father and other tappers could no longer roam over large areas of forest owned by a rubber baron; instead, the new landowners generally cut the trees on their plots, sold some timber, burned the rest, and then planted coffee, rice, and manioc. Her father and others went from struggling rubber tappers to subsistence farmers, degrading the land in the process. These clearings loosed all kinds of pestilence from the forest, and malaria overwhelmed many of the small settlements. Silva had malaria five times. Two of her sisters died from it. Her mother died of meningitis, and at fourteen years old, Silva became head of her household.

Soon, Silva succumbed to hepatitis, which, she eventually learned, was coupled with mercury poisoning. Today, legend has it that her environmentalism grew from a desire to exact revenge against the pollution that nearly killed her. For months, her health languished, and her father realized that she needed serious medical attention. She went to Rio Branco, lived with a cousin, and sought treatment at a free Catholic hospital. In the city, she came to understand the burden of illiteracy, which motivated her to take a basic course in reading and writing. She moved into a convent as a housemaid and enrolled in school.

In the late 1970s, the Catholic Church stood virtually alone in organized opposition to the military government, and Silva was influenced by the sense of social justice of the nuns around her. At this time, an activist named Wilson Pinheiro began to organize rubber tappers in Acre in acts of resistance called *empates*. The tappers would participate in sit-ins on unclaimed land, or land of dubious ownership, and claim the right to use it. These confrontations resembled the sit-ins of the American civil rights movement, often ending violently. Two *pistoleiros* murdered Pinheiro in 1980. At his funeral, "fifteen hundred of [the rubber tappers] queued up to swear vengeance, placing their hands on his corpse," wrote Alex Shoumatoff in *The World Is Burning*. The tappers stormed the ranch of the man suspected of orchestrating Pinheiro's murder, tried him on the spot, and executed him. The police reacted quickly. Hundreds of tappers were arrested, many tortured.

Pinheiro's death and the government's overreaction to the tappers' revenge inspired a movement. "The cycle of violence that began in the hot dry days of July 1980 swirled in rising currents around Acre in the months and years ahead," Andrew Revkin wrote in *The Burning Season*. These currents produced homegrown leaders with a new vision for the Amazon, a vision that brought the focus to a forgotten animal in the environment—human beings. Until Pinheiro's death, soon to be followed by the more celebrated life and death of Chico Mendes, the debate over the forest's fate rarely accounted for the fate of the millions who lived there. In the next eight years, Mendes would champion the plight of the rural poor and landless. It would cost him his life.

By the mid-1980s, Silva was a married mother of two, a university student, a schoolteacher, and an organizer for the Workers' Party, and the *empates* still occurred on a regular basis. The impact of these sit-ins spread as Silva and Mendes, with advice from Mary Allegretti (an anthropologist from Brazil's south who would become Secretary of the Amazon under Silva) and Steve Schwartzman (an American anthropologist who, after working with the Kreen-Akrore tribe in Acre, joined the Environmental Defense Fund in Washington), determined that articulating an agenda of "Save the Amazon" would bring the rubber tappers and their concerns to the attention of the international environmental movement and would hopefully bring the financial support the movement offered. This campaign became the basis of Mendes's breakthrough coalition: by appealing to an international audience concerned with trees, he sought a security blanket for his comrades—the men and women living among those trees.

A natural paradox arises from this marriage, as environmentalism traditionally opposes human occupation of the forest. Silva and her colleagues characterized their goal as "sustainable development." The term, as used in the Amazon, is intended to connote a harmonious existence between the environment and denizens engaged in nondestructive activities such as rubber tapping and nut gathering. But the term has become hollow. Since the early 1990s, the moniker has been co-opted by such a wide array of actors to justify their presence and activities that "sustainable development" is nothing more than a relative term (e.g., soy farming is more sustainable than cattle ranching), and no longer a tenet of environmentalism.

Silva was in São Paulo at the time of the Mendes murder in 1988, seeking treatment for the hepatitis that continued to plague her. As a visible survivor of the rubber tappers movement, she was swept up by a wave of empowerment born of the sympathetic reaction to Mendes's death. Silva perceived that politics offered greater opportunity for her neighbors than violent confrontations. She had befriended a union leader from São Paulo who had come to speak at Pinheiro's funeral—Lula. Along with Mendes and others, she formed a chapter of the Workers' Party in Acre, a rural ally of Lula's urban union-centered organization.

She easily won election to the state legislature in 1990 on a platform advocating sustainable development—arguing that development and destruction did not have to go hand in hand. Her service was interrupted, however, by another bout of bad health, when the mercury poisoning was finally identified. She was pregnant again, and the treatment was precarious. It took her nearly a year to recover her health.

Still, her popularity never abated, and in 1994, she was elected to the federal Senate at the age of thirty-six. Dark-skinned, poor, female, Amazonian, illiterate for almost half her life, chronically ill—she was also a national heroine. She became a leader of the Workers' Party in the Senate, and it was a foregone conclusion that she would be given the environment portfolio when Lula won election in 2002. Marina Silva's becoming Minister of the Environment symbolized that anyone could achieve anything in this country.

She spoke to us in Brasília from behind a desk that evidently had been built for a tall man with a big ego, as if the bureaucracy's furniture designers hadn't planned for the eventuality of her holding office. Her dark hair was pulled back tautly, and her face was bone thin. She sipped tea as she spoke. She could eat only a handful of foods, nothing processed. She was frail, often missing appointments because of illness, and spoke in a hard whisper.

For almost an hour she answered questions in soft, clipped sentences full of principled rhetoric. We had heard that she was overwhelmed by her job—that the nasty scum in industrial cesspools in São Paulo, the filthy air in Rio, and the relentlessly depressing deforestation statistics had taken a toll on her optimism. We had also heard that she was butting heads with the ministries of agriculture and mines and energy, which focused on export revenues more than preservation.

She acknowledged that she hadn't anticipated the collaboration needed among ministries for her agenda to succeed, that it was easier to advocate and criticize from the Senate than to try to carry out policies from within the administration, even though everyone came from the same political party. "There's not much that we can accomplish without resources, and there is a competition for resources among all the ministries," she said. She explained that her job was easier in one respect because past legislative efforts meant she didn't have to focus on enacting new environmental laws; the ones on the books were as progressive as anyone could hope for.

But she needed the means to enforce those laws, and she understood that the means might lie beyond her reach. "I know that money is an important resource, but we now have an even more important resource—an ethical effort by a government that is principled and stands for social justice. This will help us gain the respect of our population and the respect of other countries. This is intangible capital, but we need it in order to make sure we have the support for whatever it is we hope to accomplish."

To believe that principle will take the place of money is a tenuous form of governance. Yet to some extent Silva had no choice, as she was coming to learn. Her ministry was wedged in between the other ministries, which had mandates to produce gross national product numbers that provided jobs and a tax base—ministries that traditionally eschewed environmentalism as a luxury for developed countries. Lula himself had reminded environmentalists of this reality in June 2003 when he said, "This region can't be treated like it was something from another world, untouchable, in which the people don't have the right to the benefits."

It was up to Silva to spin this message to the environmental community as support for sustainable development, which it wasn't. The sine qua non for political reform, Lula was realizing, was economic success, and the Amazon was expected to do its share. Lula, who had committed his government to the Zero Hunger program, was willing to cut a tree to feed a child, or as many trees as it took. In his inaugural speech, he said, "If, by the end of my term, all Brazilians are able to eat breakfast, lunch, and dinner, I will have fulfilled my life's mission." The environment, to the extent it impeded economic development, was secondary.

Yet Silva was unwilling to espouse what was required by this social and economic initiative. A true believer that there didn't have to be a trade-off, even with depleted resources and a secondary seat at cabinet meetings, she still had a trump card: the government couldn't afford her public displeasure. As Silva's tenure with the government lengthened, she also capitalized on another asset: integrity. The Lula government succumbed to multiple resignations over allegations of corruption, but Silva's office remained untouched by these allegations. Policies in many ministries may have been up for sale to the highest bidder, but not hers. Scandals also arose in the pro-agriculture and ranching state of Mato Grosso. These scandals weakened the power of politicians who were advocating resistance to federally imposed environmental restrictions.

At the same time, Silva began to appreciate how to use her bully pulpit and goodwill by moving from the strategy of an *empate*, or standoff, to one of engagement. She began a dialogue with Governor Blairo Maggi of the state of Mato Grosso, the world's largest soybean producer, to encourage him to incorporate land use planning and watershed management into the expansion of the agricultural frontier. In her efforts to impose rational planning in the paving process of BR-163, she explained to local mayors at the periodic regional meetings she attended that law and order provided the foundation for economic growth, and encouraged them to focus on land titling issues as a primary cause of violence. She invited researchers and NGOs to these meetings that included politicians and businessmen, exposing all the participants to dissident points of view—a novel concept in a region where nonnegotiable decisions had long been imposed from afar. Her acknowledgment that economic prosperity was not inconsistent with preserving the environment meant that compromise was attainable. "When you take care of the people of the forest, you are taking care of the forest," she said.

Silva's effectiveness in the Amazon will be judged to a great extent on how the paving of BR-163 pans out. She has adroitly accelerated the set-asides of national reserves in response to the outcry over deforestation rates and the Mendes-like reaction to the murder of an American nun in the Amazon in February 2005. These efforts, however, do not require the same type of finesse as the negotiations over the road. If Silva succeeds in imposing even a veneer of rational planning for BR-163, she will have proven that a spectrum of intransigent interest groups can forge a compromise. A revolutionary concept in the Amazon.

Interviews with Silva are sought after because of her celebrity but frustrating because her rhetoric ("We need to harmonize the needs of the people with the needs of the forest," for example) remains platitudinous. Her interviews convey a kind of mysticism, explaining why many expected that she'd quickly resign in frustration from her position in the government. But her actions reveal a savvy political infighter. Handicapped by scarce resources and in competition with fellow ministers, Silva has figured out how to monetize her principles. Since she cannot buy influence by withholding or bestowing public funds, she substitutes her persona for these assets. Her triumphs, should they come, will be victories of character as much as those of a cause.

The importance of Silva's home state of Acre to the development of the Amazon is surprising, given its isolation, tucked in the western corner two thousand miles from Brasília. And given its size, about one sixth the size of Mato Grosso and an eighth of Pará. The rubber tapping industry dominated its economy until the early 1970s, when the road leading from the capital of Rio Branco to the Bolivian border was completed. The influx of immigrants upended the traditional rubber tapping economy, leading to a consolidation of landholdings; by 1982, the fifteen biggest landowners in Acre owned 26 percent of the state.

Because of the state's remoteness, a feudal oligarchy evolved as a ruling class without interference from the federal government. The disparity of wealth created social and political schisms, highlighted by the Mendes-led *empates* and by the pervasive corruption. Acre's proximity to Bolivia and Peru made it an ideal transshipment point for cocaine, a drug that poisoned the social order as much as it did its users. The two principal U.S. books about Mendes's murder—Shoumatoff's *The World Is Burning* and Revkin's *The Burning Season*—portray an amoral society in Acre divided into those bent on self-preservation through violence and those bent on social justice through violence. In such a sparsely populated area, everyone knows everyone else, and it's as if this warfare were going on between rival gangs in the same high school.

The international spotlight following Mendes's death provided the first dose of disinfectant. Brasília was loath to have the world's press assume that Acre and the rest of Brazil were interchangeable. Among the remedial measures taken by the federal government was the approval of a final action plan to pave the 312 miles of BR-364 from Porto Velho to Rio Branco, providing a reliable land link to the rest of the country. Also, the corruption got so widespread that, ironically, it remedied itself.

In 1992, Governor Edmundo Pinto was assassinated in a São Paulo hotel where he allegedly had gone to negotiate kickbacks for public contracts. His death, rumor had it, was ordered by federal congressman Hildebrando Pascoal, who was angered at his exclusion from Pinto's largesse. Pascoal had a particularly colorful career: in 1999, the Federal Chamber of Deputies voted to expel him for his role in leading a death squad responsible for the murder of 150 Acreans since the early 1980s, including one victim whose limbs were severed by a chain saw and another who was dipped in a vat of acid.

The demise of this local political mafia coincided with the emergence of the Workers' Party, led by Silva and Jorge Viana, who served as the mayor of Rio Branco before being elected governor in 1998 at the age of forty-two. Viana, a forest engineer who worked closely with Mendes, took office with a mandate to create a "Government of the Forest" and to make Acre Brazil's first "green state." To do so, he realized, he would need money in a hurry. His was a minority party in Brasília, so Viana focused elsewhere. The *Chicago Tribune* reported on Viana's success in his initial efforts, "His progressive views toward the rain forest . . . coupled with a broad political cleanup and an efficient new administration staffed by technicians rather than political appointees, already [within the first six months] has drawn $200 million in national and international funding to Acre state." (Viana pointed out to us that expectations surpassed reality; the actual number was $108 million.)

Viana's fund-raising success should have come as no surprise, given his telegenic good looks and engaging manner. "I intended that the first four years of my administration would be for planning and, if I were lucky to get reelected, the next four years would be for carrying out those plans. Money was very necessary to all of it," he told us. Leaning into the polished blond wood conference table in his office in the center of Rio Branco, Viana described his vision for Acre as a mix between Costa Rica and Finland—two countries that have made the most of their substantial forests. "Finland has a strong forest-products sector and a high standard of living. Costa Rica is a tourist paradise as well as a bio-laboratory. We think we can be both. And you can start to see the evidence."

The first "evidence" a visitor to Rio Branco sees is a state-of-the-art airport connected to the city of three hundred thousand by a smoothly paved, well-lighted road. "First impressions are important," Viana said. His devotion to public works is evident in the miles of bike paths that curlicue through the city. These paths have replaced unsightly and rank sewage troughs and have provided a means for alternative transport. Nowhere else in the Amazon is bicycle traffic so heavy. The roads are in better shape here than in most Amazon cities, and there are municipal parks with jogging trails and clean restaurants. Viana emphasized that "people need to like where they live. They take better care of it, and it makes them better citizens." Rio Branco had been plagued by blight, primarily because of its inability to handle the overnight influx of unemployed rubber tappers, reflected in the increase in the state's urban population from 47 percent in 1970 to 75 percent in 1980. It wasn't until Viana took office that the infrastructure was upgraded.

Viana's renown, however, was not for what he had done in the city but for his vision for the rain forest and how to provide economic opportunities for those living there. His model differs from the norm. Rather than opening the forest to migration and then managing settlements, he advocates incentives for those settlements already in place, such as subsidizing rubber and Brazil nut prices. Viana's vision also is to move the processes that add value closer to the source, such as constructing rural factories and processing plants. Providing job opportunities in regional rural centers, he reasons, reduces the incentive to cut the forest for logging or cattle. The state government guarantees a minimum price for rubber, and now a substantial amount of that production goes to a new condom factory in Chico Mendes's hometown of Xapuri. Roads to Bolivia have been improved to promote trafficking of Brazil nuts for processing.

"In 2002, a can of Brazil nuts was worth about five reais [about $1.50]," Valerio Gomes, a doctoral student at the federal university, told us. "Today a can is worth twenty reais, improved by greater market demand. A rural family will now be less likely to cut down an area of forest. They'll look and see that a certain area contains four or five Brazil nut trees from which they can earn so much money per year, and they'll be less likely to burn the area or sell the wood." The decision of whether to deforest becomes a choice between two balance sheets. Viana's goal is to present forest residents with alternative economic uses of their land. He wants them to compare the present stream of income from harvesting Brazil nuts with the potential income from logging the same area, planting it for pasture, or raising cattle on it. He has encouraged NGOs and researchers from the federal university system to assist Acreans in weighing these choices.

Acre has also become a staging ground for the development of certified forests, where extraction takes place in an area according to a predetermined cycle, with reseeding following logging. Trees are cut strategically to avoid damaging their surroundings, and extraction of tree species is orderly to avoid depleting local diversity. As part of his add-value-in-place program, Viana subsidized the construction of a furniture factory in Xapuri, which produces high-end products sold in São Paulo with the cachet of certified wood. "It's a way of involving consumers in making environmental decisions," Viana explained. "If they will only buy certified wood, then certified forests that provide good wages and safe conditions for the people and the environment will do well."

Yet the Government of the Forest still comes down to a matter of dollars and cents. Will consumers pay a premium for products as a way of participating in saving the rain forest? Will the advisers Viana sends into the forest be able to convince the tappers and nut growers that a ten-year plan based on extraction of products is more desirable than a two-year plan based on logging and cattle ranching? The governor of the neighboring state of Rondônia, often the butt of Acre's criticism on how development should not be done, has opined, "The state of Acre is a place with excellent ideas and speeches and without economic results." Beleaguered by the cross-border derision, the governor of Rondônia warned, "If the government of Acre doesn't stop criticizing Rondônia, I'm going to close the national highway, and the people of Acre are going to die of hunger."

Acre's isolation allows it to entertain policies that larger, more populated states might consider dilettantish. The state's isolation also highlights the artificiality of Viana's success. Acre is not a self-sufficient state. The environmentally friendly industries, such as the extractivist reserves, are heavily dependent on out-of-state funds to subsidize industries that couldn't stand on their own in Acre or elsewhere. And these are not industries of significant employment, so one must wonder if the public money could benefit more elsewhere.

Viana acknowledges the importance of subsidies, but he argues that subsidies are intended to encourage economically desirable activity everywhere, be it cotton farming in Alabama, wheat harvesting in France, or nut growing in Acre. That a government promotes certain industries instead of others shouldn't be controversial. Government's purpose is to encourage citizens to make better choices, which means spending public money to encourage these better choices. Hiring teachers, doctors, and economists is standard government practice; their allocation is a political choice to promote particular policies. Viana hasn't restructured Acre's government, just its priorities.

The effectiveness of Viana's vision depends not so much on its perspicacity but on its durability. Unless Viana can institutionalize the physical improvements of the Government of the Forest and embed his principles in his constituency, his way of governing runs the risk of being dismantled as quickly as it was assembled. Viana's term ended in 2006. The corrupt politicians awaited a change of administrations, allegedly taking orders from the imprisoned congressman "Chain Saw" Pascoal. When Viana ran for reelection in 2002, his opponents temporarily disqualified him by convincing the local electoral board that his campaign's use of the logo of a stylized Brazil nut tree, also the logo of the Government of the Forest, was a personal misappropriation of a government asset. That decision, overturned shortly thereafter, was rendered by officials who refused Viana access to briefs prior to the hearing and who denied his request to present evidence and witnesses. The governor's enemies stand ready to return.

The legacy of the Government of the Forest also faces economic and cultural challenges, no matter who succeeds Viana. If these industries do not become self-sufficient over time, then their benefactors will eventually turn away, not necessarily because of the continuing cost but because of the projects' diminishing precedential value. International donors, convinced that this model cannot be replicated, will move resources to places and causes where economic experiments justify themselves over time and prove transferable to other areas with similar problems, a sort of "prove it or lose it" proposition. These projects also face waning indigenous support, a question of the sustainability of sustainable development. The doctoral student Gomes points out that the next generation "has a wider array of knowledge to draw on to decide what's good and what's bad for them." Children of the forest now have an appreciation of what city life has to offer in terms of education, entertainment, health services, and social interaction. The childhood of Marina Silva, as abject as it appears, is typical in the remote areas of the forest. Without a next generation committed to the economic way of life promoted by the Government of the Forest, it could soon be a relic.

Irving Foster "Butch" Brown awakens in Rio Branco at four A.M. to prepare for one of his frequent visits to the center of the universe. It's his favorite time of day, when the human cacophony of his neighborhood sleeps and the noises of the remaining wildlife bordering his cluttered *bairro* remind him of his surroundings. Brown, a mid-fifties professor at the Federal University of Acre where he specializes in land use, enjoys the moments the quiet gives him to prepare his lessons. With a strong Brazilian coffee alongside his laptop, he sends scores of e-mails to his classmates from Amherst College, his scientist colleagues in Woods Hole, Massachusetts, and his contacts in the world's press and NGOs. He writes to alert and inform them of the consequences of the proposed TransPacific Highway that will link the Atlantic to the Pacific: the threats the highway poses to the environment as well as the upheaval it will cause to Amazonian ways of life. He's been in Brazil for more than twenty years. In the 1990s and early 2000s, Brown provided his correspondents with periodic installments of *Innocence in Brazil,* his encounters with kidnappers, snakes, interminably bad bus rides, and several life-threatening diseases. Lately, he hasn't had time for these exploits.

"The road is coming," he says. "And we're not ready. We're fighting time to get the people ready. We're trying to build coalitions to help the area develop, while preserving the land and water people need to survive. And too much of my time is spent trying to raise money to support all of this." The bureaucratic part maddens him.

Several coworkers straggle into his cluttered stucco house near the university, which he shares with his longtime girlfriend Vera Reis, who has spent several years developing educational programs for the children and adults in the border town of Assis Brasil to prepare them for the highway—the increased traffic, the deforestation, the influx of immigrants, and the arrival of large-scale commerce. Brown and Reis have commuted regularly to the border over the past five years, forecasting to the local population about the impending change.

Brown's bailiwick is along the Road of the Pacific, a transcontinental road connecting Rio de Janeiro to the Pacific and opening another vein into the Amazon, particularly into the environmentally delicate Madre de Dios area of Peru. The dream of connecting the oceans is centuries old in South America. The road will pass through three countries, each at a different stage of development and thus each with different priorities. The Amazon constitutes half of Peru, although only 5 percent of its population lives there, so it has escaped major deforestation—so far. Roads through the rain forest in Peru will have the same impact they've had in Brazil. Within Peru itself there's still debate as to what port will serve as the terminus and reap the income from the increased traffic, although Ilo is the front-runner. Losing the Madre de Dios region to deforestation should cause regional climate change, if what's happened in Brazil is any guide. Bolivia recently experienced nationwide riots protesting the internationalization of its gas resources, a sign of concern about Brazil's dominance. Yet, the road promises Bolivia an end to its landlocked status, which once prompted Queen Victoria, peeved at its treatment of her ambassador, to declare that a country without a port simply "does not exist."

Brown has seen the ramifications of roads—their invitation to illegal loggers, slash and burn settlers, and real estate speculators—and the impact on local populations. Feverishly, as if he were running ahead of flowing lava, he has promoted fieldwork among his graduate students to encourage them to spread literacy, hygiene education, and rudimentary surveying skills so that local populations will understand what they own. To complement these efforts, he has organized regional meetings within Brazil and has coordinated colleagues across the borders under the MAP banner—M(adre de Dios, Peru), A(cre, Brazil), and P(ando, Bolivia). He's become a sort of folk hero to the locals and his students, not only because of his obvious selflessness but also because of his permanence. Many researchers come and go; Brown is a fixture. "To most people, this is the end of the world," he says. "Not to me. I live here. To me, this is the center of the universe."

Monica de los Rios, Elsa Mendoza, and Nara Pantoja show up in the predawn hours, each with a backpack containing water, a small knife, a first aid kit, pens, a notebook, and a laptop. Brown has packed the Jeep with hammocks, rope, machetes, a large first aid kit, and a wooden stretcher. They decide on a driving schedule and take off before first light.

The air has a chill at this hour, but the smokiness still stings, as the dry season has again brought its share of fires. There's anecdotal evidence of increasing incidences of asthma along the main roads in Acre, which is easy to believe, given the taste of the air. The rivers were especially low in the summer of 2005, as if struck by drought. Acre is one of the places where weather dominates conversations. Here it's seen as evidence of human activities, not just the product of supernatural forces. Too much burning, not enough rain. Too much flooding, not enough forest cover.

Passing over the Acre River, we see the *gamileira,* the oldest road in the city. It has been revitalized under Viana's administration and is bordered by buildings painted pink, yellow, blue, and peach—a fruit salad cityscape. There is a *danceteria* in the neighborhood, packed on Saturday nights with the *forro*-dancing crowds (a modern country version of formal European quadrilles). As we leave the city, we see the site of the new soccer stadium across from the strip of stores with farming supplies and auto parts. We're on the Chico Mendes Highway. This highway connects at a traffic circle to BR-364, which heads toward the farthest west city, Cruzeiro do Sul, and east to Porto Velho and the Road of the Pacific, our course on this day.

The landscape soon turns from small weekend homes to large cattle farms, miles and miles of pasture on each side with trees barely visible on the horizon. Brown points out that what we're seeing is not the Amazon but Africa (the grass varieties) and India (the cattle breeds). In the rainy season, he says, the pastures are green and smoky from the evaporation. Now they're brown and yellow and smoky from fire. Without the enclosure of forest, the sky is high and deep, the clouds full and slow-moving. The road curves and slopes to maximize the high ground. Low points wash away in the rainy season. Brown stops from time to time, making notes of cracked pavement. He'll call a state official on his return, or complain in a letter to the newspaper. He's a gadfly.

The good pavement wanes at Senador Guiomard, a small town about forty minutes from Rio Branco, just a *churrascaria,* a bus station, an outpost of the Bank of Brazil (along with a priest, the first sign of civilization along Brazil's frontier), and a half dozen speed bumps. In the small towns we pass, we see as many ox-drawn carts as cars. The street bustles at six A.M., especially around the bread stores offering free shots of espresso. The next sign of life is a military police station, uninterested in a vehicle heading *toward* the border. Brown points it out as one of many signs of Brazil's concern about opening a border with Peru and Bolivia, especially so proximate to coca-leaf-growing regions.

The fabled Xapuri, where Chico Mendes was gunned down, lies an hour and a half from Rio Branco, several miles off the main highway and worthy of a visit. Julio Barbosa, the mayor, rides up to city hall on his bike: fat tires, red frame, no gears. "I'm sorry I'm late," he says. "I had to go to a funeral that just went on and on." He pauses. "And first you must excuse me, so I can have an interview with a radio station in

Peru. They want to know what they're going to see when the road opens up."

Barbosa is another luminary from the era of Mendes, his successor as president of the rubber tappers' union. His constituency is a town of seven thousand people that has become something of a national showpiece, and the beneficiary of government investment, evident in the new school and the condom and furniture factories. Mendes's baby blue pink-trimmed house still stands as it was on the day he died, surely one of the most remote museums in the world. The sturdy stucco buildings from the rubber boom line a tidy main street. A procession of ox-drawn carts clatters down the center interspersed with a few pickup trucks. In the middle of one of the side streets, a big yellow dog tries to fight his way through a flock of buzzards to get a piece of discarded meat.

"Life here is much better than when Chico was alive," Barbosa says, "thanks to God." But he complains that the town's industry rests on a slender reed. The factories depend on subsidies—for the certified wood or rubber, or through government support for the nut factory. He knows that if the subsidies were taken away, the factories would shut down and people would be left to their own devices in the forest.

"That will be bad again," he says. "The native forest does not have enough monetary value as forest, so people will turn to cattle or move to the city, and those who come after them will destroy it." As if to prove the point, in September 2005, Xapuri's vice mayor, João Batista da Silva, was arrested for starting illegal fires alongside the Chico Mendes Extractivist Reserve.

During the two more hours to Epitaciolândia and Brasileia, we see some wooden shacks along the way, some signs with the names of cattle ranches, a few evangelical churches, and no trees. We come upon the "curve of death," a sharp turn in the road with a warning sign that many interpret as a challenge to speed up, unsurprising in a country where pictures of the late Ayrton Senna are as popular on walls as those of Jesus Christ. We pass a new sports center, and then small storefronts begin to appear, interspersed with houses at first. Then there's just a solid line of commerce—supermarkets, banks, general stores, bakeries, government offices. The road divides at a church with a small park. We're near the last federal police point where we'd be stopped if we were planning to take a trunk road to cross the border into Peru or Bolivia. Instead, we wait at a one-lane bridge to cross into the town of Brasileia. Donkeys swish their tails in front of us, waiting to pull their loads of rice over the bridge that can handle only one line of vehicles at a time.

Outside of Brasileia, we see our first sign of *Estrada do Pacifico* (Road of the Pacific), and the pavement improves. Signs warn of erosion areas and upcoming curves. We see wooden shacks and a few people waiting for buses with boxes of live chickens, pots, and pans. Engineering change outpaces cultural change. Brown points out these contradictory images. "It's exciting, straddling these two worlds. Sometimes it's overwhelming. At least I know I'll never be the victim of an existential crisis. I look around, and I know I'm needed. At least I think I am. There are problems plaguing this area—wildfires, diminishing water supply, an undereducated population, and massive erosion." He pauses. "These problems will be here with me or without me, that's not what I'm saying. Just maybe I'm helping people cope with them for themselves."

Just one more bad stretch of road ten minutes outside of Assis Brasil. Then we see the tri-national border marker and the road sign that says the Pacific Ocean is 913 miles away. In the opposite direction, a sign notes: SÃO PAULO 3,200 KMS (1,988 miles).

Jerry Correia, nineteen, the assistant to the mayor, is our welcoming party. He hugs Brown, tells him the whereabouts of a half dozen people he says he needs to see. Turning to us, he explains where we have arrived. "For Brazilians, Acre is the end of the world. When you say about somebody, 'He went to Acre,' you mean that he died. But for Acreans, Assis Brasil is the end of the world. So, it's the end of the end of the world."

The skeleton of a bridge across the trickling Acre River into Peru symbolizes imminent change. It's possible to wade into the muddy water during the dry season and have a body part in three different countries. A palpable sense of apprehension marks our conversations. "I think it's the drugs that will be the biggest change," Correia says. "When the traffic starts coming through, the big trucks from Peru, there's going to be a lot of cocaine moving. That's what's going to change people's lives."

Brown adds his laundry list of environmental concerns, noting that the health of the Acre River basin has been suffering, as its deforestation has robbed it of watershed protection, and the increased population has polluted its tributaries. "Sometimes you wonder if any good can come of this," he muses. "Then you wonder how you can hold this back, and you conclude you can't. So you deal with it the best you can."

Acre's business community sees the road as opening significant new markets, especially Peruvian beef consumers on the eastern side of the Andes. Soy farmers in Mato Grosso and southern Acre see an outlet to the Pacific and a port to Asia, although it's difficult to justify the economics of truck transportation over the Andes, even if the distances are shorter than for the barges and grain carriers that ply the Amazon waterways.

Dreams of interconnections among these countries don't stop at this border. Several projects are under way to connect the waterways from the Amazon/Orinoco down to the Río de la Plata in Argentina. President Chavez of Venezuela has proposed a gas line from his country to Argentina with its course through the heart of the Amazon. Mercosul, the common market of South America, is handicapped by physical barriers that could be overcome only by massive expenditures of public moneys. Is this dream worth fulfilling? Are the consequences worth the cost—the degradation of the environment, the diversion of resources from other programs, the disruptions to local communities? The Road of the Pacific will provide the first answers.

CHAPTER 12

OPENING THE RAIN FOREST

—

In the town of São Feliz do Xingu, in the southwest quadrant of the state of Pará, we visited Marco Antonio Pimenta, the local volunteer for Greenpeace—the giant international environmental defense organization that has made a particular cause of trying to control the logging of Amazon mahogany. In a series of well-documented reports, Greenpeace has made a case that the mahogany trade is as illicit as narco-trafficking. Because some of those with guns, money, and political power in the Amazon often value mahogany trees more than fellow human beings, people die defending the forest. A tree often is worth a life, especially if the tree is mahogany.

Cutting a mahogany tree in Brazil is now as illegal as growing coca, but the value of mahogany is so great that restrictions are seen as challenges to be circumvented, not obeyed. Greenpeace argues that the indiscriminate extraction of mahogany not only is depleting a rare natural resource but is also undermining the social and political fabric of the region.

"Mahogany corrupts," said Pimenta, who, in addition to his work for Greenpeace, works full-time as a security guard at the local bank.

"Most of the enforcers from the government are corrupt. They take money to allow extraction of trees on indigenous reserves. They tell the public that these are protected lands, but that is not true." He claims that licenses are issued for the legal extraction of trees other than mahogany, but loggers use the access to cut mahogany, and there is often no follow-up enforcement when bribes are paid.

According to Pimenta, IBAMA, the national environmental protection agency, seized fifteen thousand cubic meters of mahogany in Pará in 2003, a fraction of the actual extraction. Pimenta spends most of his free time preparing "denunciations," which identify suspicious activity. He sends these to state government officials in Belém or to the federal government in Brasília. "Then I wait for a reply." He paused and made the traditional Brazilian snapping sound by pressing his middle finger and thumb and leaving his index finger limp as he repeatedly flicked his wrist. "And I wait."

Pimenta, short with the build of a weight lifter, told us that mahogany corrupted the entire civic establishment of São Feliz. "The mayor's involved in a scheme to extract trees from the Kayapo indigenous areas. He has twenty-one cases against him in court. I've sent denunciations with details to Brasília. But nothing gets done. No one is going to help me."

Asked why he didn't go to the police, he snorted. "There are leaks in the federal police offices that oversee the area. When a team comes, people in Xinguara and Tucuma call to warn the loggers first, and they hide all the wood before the police arrive."

He even blames the clergy. "Padre Danilo used to talk about the conflicts, even at weddings. But not the new one, Padre Paulo. Padre Paulo has sold land—even the *grileiros* [land grabbers] say that. Everyone is involved."

From the time of the first European explorers, the big-leaf mahogany tree, which grows to more than 125 feet, has been prized for its distinctive reddish brown color, its even grain, and its long straight boards. It's called "green gold" in the Amazon. Just a slab off the end of one of those boards will make a first-rate dining room table, the kind you might see in a Manhattan showroom listed for about fifteen thousand dollars. A single tree will yield tens of thousands of dollars of furniture, paneling, guitar bodies, or even elegant coffins. Woodworkers love the wood's strength and manageable weight. Unlike some of the dense tropical hardwoods, mahogany cuts and mills with precision, then holds its shape. Shipbuilders have long appreciated that the stout waterproof planks have long, bendable spans but could shake off the impact of a cannonball just as well as the more ungainly oak. The finest eighteenth-century furniture came from a dark Caribbean mahogany, now extinct, that American craftsmen of Chippendale and Hepplewhite assembled into elegant cabinets and desks with intricately turned details and highly polished surfaces. The most common variety now is the Honduran mahogany, though most of these trees come from the Amazon in Brazil, Bolivia, and Peru.

The mahogany tree is unexceptional at first glance, with buttresses at the base like many Amazon trees and a scaly gray bark. The top, which extends to the highest layer of the forest canopy, is surprisingly small. Skilled men with two chain saws can take one down in less than an hour. It takes longer if they try to guide its fall to provide minimal damage to the forest, a rare act. As most of the cutting is either flatly illegal or on protected land, speed is paramount. The fresh wood beneath the gray bark is a reddish pink that resembles raw meat.

Access to this tree has been a significant factor in the expansion of the Amazon frontier. There is no government presence in the remote, uninhabited areas where it grows, and the loggers themselves cut the roads, a service traditionally provided by a government. The government arrives only when a critical population mass has been reached. Until then, the law of the jungle prevails—if someone with hired gunmen wants the land you occupy, you'd better leave, or you may end up dead. The government's eventual presence doesn't always change this.

Like most flora in the Amazon, a mahogany tree grows singularly. Even in the area of highest concentration in parts of the state of Pará, there are about three trees per hectare, surrounded by hundreds of other tree species, each different from the next. The loggers know where the mahogany are, though the trees seem to be increasingly rare in the heavily trafficked "mahogany belt" that runs from the eastern Amazon in a southwesterly swath. The density here before logging was one tree per five to twenty hectares. Many of the remaining trees are in restricted national forests or on Indian reserves and are forbidden to be cut. This inaccessibility does not stop loggers. There are regular reports of bribes to federal officials to condone logging in protected areas, or deals made with Indians to invade lands in return for payoffs.

The typical logger is a small-time contractor with a handful of laborers working for him. In the early morning hours he and his crew cut their way through the forest until they find a mahogany tree. Sometimes they must make their own trail, but often they follow a route tramped down by land-starved settlers or gold miners, who often blaze the trails in the Amazon.

If the logger has the means to bring a bulldozer into the site, the whole tree is usually hauled to the nearest place a truck can access. Otherwise, it's cut into twenty-foot chunks and pulled out by men working with ropes. For their efforts, the laborers might get the equivalent of fifteen or twenty dollars a day, a good wage in the Amazon.

The mahogany follows a chain of ascending value. In a 2002 article in *Outside* magazine, Patrick Symmes detailed how the tree increases in value as it goes from the Amazon to the dining room. The tree itself is free. The initial cost is what it takes to get to the tree: building a road, bribing an official, or even killing someone—or all three. Once ready for transport out of the jungle, the tree is worth $30. At the sawmill site, usually in a small town with access to truck routes, the tree is worth $3,000. The sawmill sells it to an exporter who hauls it to a port, usually Belém, and sells it to a trading company or direct purchaser for upwards of $50,000. The final retail value of the tree, once it has been turned into furniture, is over $250,000. The trade is a classic study of a third world commodity that pays little to those who extract it but achieves full value at the highest end of the first world consumer pyramid. This economy mirrors that of cocaine.

The loss of these trees is not the only part that threatens the forest. In addition to disrupting the mahogany tree's vital role in the ecosystem, extraction creates destructive pathways that cause further harm. The scars left behind by the invading loggers do not heal. These tiny trails are often visible from the air, their pattern resembling a river watershed in reverse. The end of the line is the tiny white vein that stops at the base of what was once a mahogany tree. The massive trunk is dragged back to a more established trail, which in turn leads to a solitary mud road. Working backward along that road, as it gets wider, you can see side roads jutting out like the bones of a fish from the spine, narrow at first then widening as you go back down the road closer to the presence of civilization. At the tips of the bones are square patches of farms and, farther along, vast clearings for cattle or for mechanized-agriculture crops, such as soybeans or rice. Finally, somewhere along the route is a small town.

That pattern tells the story of Amazonian deforestation over the past thirty years. Greenpeace has argued that cutting a mahogany tree and making that skid trail is the beginning of the end for a piece of forest, an irrefutable claim. "Mahogany starts a cycle of destruction," said Scott Paul, director of Greenpeace's forest program in Washington, D.C. "The tree is worth enough to fund those first cuts into the forest, and once that happens, there is so much pent-up demand that the rest is unstoppable."

The skid trail is an invitation—a port of entry for other loggers seeking less valuable but more numerous species such as types of cedar, jatoba, and the rock-hard ipe, all of which are increasingly valuable. Once the trails are cut, squatters inevitably move in, itinerant families carving out a subsistence living along the frontier. They move on when threatened or paid to leave. Activists who encourage small farmers to stay on their land often end up dead. The Pastoral Land Commission, affiliated with the Catholic Church, estimates that 475 activists were murdered in the state of Pará from 1996 to 2001. Their offenses primarily consisted of organizing the occupiers of small plots of land in challenges to the validity of land titles claimed by loggers, land speculators, or ranchers.

The general pattern, which worries groups such as Greenpeace, is that these families are supplanted by loggers, who then have easier access and are intent on consolidating larger tracts of land, clearing them, and then selling them off to land speculators. The speculators come in once the land has been cleared, and they consolidate the holdings into even larger tracts, which then are sold off to cattle ranchers. And when the soil is fertile and flat, such as in large areas of southern Pará and the state of Mato Grosso, the cattle ranches eventually give way to large agricultural plantations.

Greenpeace issued a report in 2001 concluding, "Fueled by high international market demand, mahogany is driving the destruction of the rainforest of the Brazilian Amazon. . . . The trade in illegal mahogany is just the tip of the iceberg. It signals the failure of world governments to act to protect the Amazon."

Because of its majesty and its pure economic value, mahogany is a simple way to explain a complicated problem. The Brazilian government, reacting to international pressure, has restricted the sale of mahogany until nondestructive logging methods can be developed. In 1996, the government imposed a moratorium on all new logging permits for the tree. That proved ineffective, and following the Greenpeace report in 2001, there has been a ban on all mahogany harvest, transport, and export. The tree is now covered by the rules of the Convention on International Trade in Endangered Species (CITES), which makes it nearly impossible to obtain an export permit. The government coupled this action with highly publicized raids on sawmills in Pará and confiscated 7,165 cubic meters of mahogany—valued at approximately seven million dollars.

The coordination between the government and Greenpeace could easily be interpreted as a concession by the government of its failure to control this problem. Greenpeace spent two years preparing a report called "Partners in Mahogany Crime" that identifies the chieftains of the illegal tree trade, the locations of their sawmills, and the names of the corporate end-purchasers of the contraband. The report detailed the loggers' techniques—trespassing on Indian lands, falsifying land titles and logging permits, lying about inventories of trees and the numbers of extracted trees.

The effectiveness of Greenpeace's efforts might have sparked some broader discussion about privatizing the enforcement of environmental laws. None developed. This episode, like many others, showed that committed environmentalists within the NGOs were outpacing their government counterparts in their ability to research and identify specific unlawful behavior. The public institutions protecting the environment traditionally have been plagued by charges of corruption or patronage for incompetent bureaucrats, and even when they do function, they are underfunded. Granting enforcement powers even to a Brazilian NGO would probably be a powerful fund-raising tool, through either private sources or international financing organizations. A by-product would be that government resources and personnel would be free to concentrate in other areas. Most likely, it'll never happen. Privatizing the enforcement of laws would be seen in Brazil as a loss of sovereignty, and attention would thereby be deflected from the substance of the endeavor. In addition, because of the rampant violence endemic to this problem, shifting enforcement authority would be tantamount to putting bull's-eyes on the backs of Greenpeace workers or those of any other NGO.

In any event, Greenpeace motivated the government to follow up, at least rhetorically. In the glow of the 2001 enforcement efforts, the head of IBAMA bragged, "The government is clear—we want to put an end to the extraction of illegal mahogany." The conflict between these words and mahogany's allure showed up in headlines in June 2005, when eighty-nine people were arrested and charged in a scheme that had resulted in illegal deforestation of $370 million of Amazon timber since 1990. The head of IBAMA in the state of Mato Grosso and forty IBAMA officials were among those arrested and charged with falsifying logging permits and allowing logging on forty-eight thousand hectares of protected lands.

To understand how these forces play out over time, we returned to the town of Rendencão in the south of the state of Pará. Twenty-five years ago, Rendencão had the reputation as the most violent town in the Amazon. The immediate cause of violence at that time was the discovery of gold nearby. As the closest settlement to these mines, Rendencão became a center of commerce. Dozens of planes flew in and out of the town even though there was no airport. The main street was used for buses, cars, and planes. Everyone carried a gun, cowboy-style on the hip, and murders occurred once a day on average.

At the time, we interviewed Colonel Ary Santos, the undercover federal marshal sent by Brasília to quell this violence. As soon as he arrived in town, a man staggered toward him. Santos figured the man to be a drunkard coming to shake hands with a newcomer. But then he saw a long fish-scaling knife sticking out of the man's back.

"No one did anything. I couldn't believe what was happening. He wasn't dead yet, but no one moved to help him. If I did something, they would have known that I was from another world. The police were part of the problem. They were corrupt. They were killing for money and looking the other way for money. There was no order," he said.

He knew for whom to look. He had compiled a list of twelve *pistoleiros* believed responsible for many of the killings. There was Pedro Parana, known by his crooked nose once sliced by a knife, and Zezinho de Condespar, who ran a murder agency. For fifty dollars they'd kill a nobody; the price went up in step with the importance of the victim.

Santos organized a secret meeting with leaders of the town—the doctor, the merchants, and the ranchers. Everyone wanted Rendencão cleaned up, and they would do what they could to help. A week later,

Santos descended on the town with sixty national security agents in two bulky Buffalo cargo planes. Within twenty-four hours, they captured eleven of the twelve gunmen on his list. "We held all of them in jail for two to three months," Santos said, "but nobody would testify against them. They were afraid. So we had to let them go. But I told them clearly if they ever came back, I would be there waiting for them."

Rendencão grew from that dusty chaotic town to a thriving commercial center of seventy thousand. The violence has declined markedly. Its name in English, "Redemption," fits. There's no evidence that the rain forest ever thrived there, other than the presence of the heat and the rainy season. There are no trees for miles around.

Mariano and Carmelinda Sroczynski, children of immigrants who came from Poland and settled in the southern state of Rio Grande do Sul, immigrated to Rendencão in 1975 and are among the founding families.

"I came here to test out the area as a base for logging operations," said Mariano, sixty-six, whose blue golf shirt, dark pants, and black thick-rimmed glasses made him look more like a Florida golfer than a Brazilian logger. "It wasn't even a town yet." His wife, dressed in flamboyant fluorescent blue shorts and a sleeveless white T-shirt, and with fire-engine-red fingernails, was eager to tell her story.

"That's right," she said. "There were no stores. There wasn't even a 'weekend.' There was no Christmas. We brought the first Christmas tree to the area. We even had to teach people how to celebrate birthdays. We showed them how to make birthday cakes. Can you imagine? There were no lights. No refrigerators, no cars, no real roads. Back then, the only transportation was by plane, and they took off where the town square is now." Their house is air-conditioned, and cell phone service and broadband Internet are available throughout the town.

The Sroczynskis established a logging business focusing on mahogany before there was any government effort to control the trade. "The market was exploding then," said Mariano, explaining the relative comfort in which they live.

No longer, they said. The nearby trees have all been cut. IBAMA's regulations and other government bureaucracy stifled the trade by being inefficient, discouraging legitimate companies from doing business. "It doesn't stop logging; it just stops legal logging," Mariano said. As with so many of the rules in the Amazon, unregulated chaos or unbearable controls can be circumvented by those who know how to game the system. "There used to be twenty logging companies just in Rendencão, but almost all of them are closed," Mariano said. "And still there are many trees cut."

Carmelinda added, "When there was plenty of mahogany, we never had a problem. But now our business needs more infrastructure and support, and we're not getting it."

Logging has diminished so much that their twenty-three-year-old son signed on in 2004 with a logging company sending him to the Congo. According to Carmelinda, that's where the logging money is now—in Africa. "Not here. There's the cost of transporting the raw materials from two hundred kilometers [125 miles] away to here. That's high. The cost of petroleum is very high now. Trucks are expensive. There are taxes. There are no incentives." These problems came with the trade, but once the loggers lost their highly profitable mahogany, the other side of the ledger began to overshadow their business.

"But the real problem," she said, as if she wanted to slap someone's hand with a ruler, "is the government. To get approvals, you have to go into eight to ten offices, pay for mapping the land, for making up the demarcations. You have to travel all the way up to Belém and wait there for a week in a hotel or who knows how long, all the time incurring expenses. You're losing time and money, not working, waiting for nothing. Nothing comes out! IBAMA's demands impose costs greater than the profits, and the companies suffer losses."

According to her, IBAMA has it all wrong. It's the farmer, not the logger, who needs to be regulated. "People blame us for deforestation. The logger is supposed to be the villain. But look, there are hundreds of little trees under the one big tree. When it falls, it makes a sunny spot and the little ones grow. I take out one, but leave ten. They grow right up and fill in the forest. I don't want the forest to disappear any more than other people who depend on it for their livelihood."

But the forest is disappearing, and the loggers are the responsible parties, at least initially. Mahogany's mystique makes it a readily identifiable case study in how the lure of a freely available commodity initiates a cycle of destruction. The ban on logging mahogany may, in official statistics, result in a diminution of its extraction, but worldwide demand grows independently of Brazil's desire to protect a particular species. There will always be a market for tropical wood. Unless logging controls are imposed by a system more effective than an ineffective bureaucracy, the result will be uncontrolled logging. Putting the Sroczynskis out of business by imposing institutional costs has not dampened the market for the product.

Once loggers extract even a single tree, that area inevitably loses its natural protection forever. The access the loggers provide becomes the spigot for a flow of immigrants who will occupy the land in predictable cycles, starting with small farmers who will clear more land, giving way to small cattle ranchers, then to larger ranchers, then, depending on the agronomy, to large mechanized agriculture. This cycle of occupation reverberates again and again throughout the developing Amazon.

CHAPTER 13

AND THEN COME THE COWS

—

Maraba, the traditional hub for logging and gold mining activities, also serves as the capital of cattle ranching in Pará. Located in southern Pará at the intersection of trunk roads of the Belém-Brasília and TransAmazon highways, Maraba is crossed by the railroad that runs from the mineral-rich Carajas hills to the port of São Luís and by the Tocantins River, which is navigable all the way to Belém.

Despite its ideal location as a transportation network, Maraba sits on an active floodplain, and the city had to replant itself as it grew. The first settlement, built low along the river, flooded out nearly every year, forcing the residents to flee for weeks at a time. After a while, they simply stayed on higher ground and built a town called New City. Then, when the TransAmazon Highway arrived in the 1970s, New Maraba was built to house the government buildings and resettlement projects.

Because the cattle ranching industry matured first around Maraba, many of the early conclusions about ranching's impact were drawn from nearby areas. During our first series of visits, we traveled uncomfortably by bus from Maraba to Belém for more than eighteen hours and saw evidence of the herds. We wrote then, "We gazed out the window at evidence for dire predictions of what would happen to the entire Amazon if it were cleared. The stampede into the jungle here had begun only two decades before with the completion of the Belém-Brasília Highway. Rapidly, trees had been cut and burned to make way for pasture. Crops of grass sown in its place had grown sparser each year until it quit growing altogether. Thin cows wandered miles between meals. The desert of failed ranches went on for an hour—gray sky, gray dirt. It was 50 miles of moonscape."

The lunacy of the activity became even more clear as we traveled from ranch to ranch and found out that no one was making money. They had perfected methods of deforestation but were clueless on how to raise cattle profitably.

The cattle industry has been under attack since the environmental movement first focused on the Amazon. In the 1970s and 1980s, the government offered generous subsidies throughout the country for cattle ranching, estimated at just under $3.2 billion in 1990 dollars. The government's motivation was based on geopolitics, not economics: to provide for Brazilian occupation of large areas of land in the Brazilian Amazon. Typically, the land grants and financial support went to well-heeled individuals or large corporations cozy with the government.

In 1989, Susanna Hecht and Alexander Cockburn observed in *The Fate of the Forest*, "Livestock became the definitive land use of the Brazilian Amazon, occupying more than 85 percent of the area cleared. At very low levels of productivity, one animal per 2.4 acres, or one hectare, the costs of raising cattle were rarely met by the selling price. On the other hand, the value of the subsidized credits [and tax breaks] and soaring land values generously compensated for the risible production performance of the cattle."

The furor over this subsidized destruction intensified in the 1980s when a sensational, but inaccurate, rumor spread that "the hamburger connection" was the primary culprit of Amazon deforestation. The Internet is rife with stories from that time that "McDonald's has been clearing the Amazon for 20 years." People in the Amazon, the theories went, were landless so that Americans could eat the cows that displaced them. None of this was true (at the time, the United States wasn't importing Brazilian beef), but it did make good press.

Yet the premise of the attack on cattle ranching appeared valid: "Several studies have shown that livestock in the Amazon is not profitable without subsidies or speculation," wrote Hecht and Cockburn. The strategy of the environmental movement, thus, could be relatively simple: if they cut off government subsidies, cattle ranching would cease. The forest would be safe. Part of the plan was implemented, and the government curbed subsidies by the early 1990s. However, from 1990 to 2005, the cattle population in the Amazon increased from 26.2 million head to 65 million. At the same time, the total amount of deforested land in the Amazon increased from about 41.5 million hectares to 62.5 million hectares. With a 5.5 to 1 ratio of clearing for cattle versus clearing for farming, the cows were the culprit.

Somewhere along the way, a significant assumption underlying the "stop subsidies and you will stop cattle" argument changed. Cattle ranching became profitable. A breakthrough study conducted by Sergio Margulis of the World Bank in late 2003 highlighted this change in economics: "[Beef] cattle ranching in Eastern Amazonia or on the consolidated frontier is highly profitable from the private perspective and it produces rates of economic return higher than those obtained from the same activity in the country's traditional cattle ranching areas [in the south]. In addition to the availability of cheap land, these returns are the result of surprisingly favorable production conditions—mainly rainfall, temperature, air humidity, and types of available pasture. The direct return on cattle ranching itself (excluding profits from the sale of timber) consistently exceeds ten percent." The business is even more impressive when you consider that ranching land is an appreciating asset. The land's appreciation by itself offered a private alternative to government support, as banks, viewing the land as adequate collateral, willingly lent capital to support cattle ranching. Now the operation and the asset underlying it make money.

These trickle-up economics frighten preservationists. They now must confront the Coase Theorem, named after the economics Nobel Prize winner Ronald Coase: people will use resources in the way that produces the most value. If logging is profitable, then logging will take place. But if the cleared land is used for unprofitable cattle ranching, then expansion at this level will stagnate. Without government subsidies, the cattle industry could not survive. In this scenario, not only would people in the Amazon abandon cattle ranching, but there would also be a trickle down to logging. There wouldn't be ready buyers for cleared land, which would reduce the incentive to do so. That was the expectation when subsidies were eliminated. But as the upstream cattle enterprise became profitable, that motivated the suppliers of the raw product—land, in this case—to supply more inventory. The change in economy threatens ecology.

We traveled to the fairgrounds on the outskirts of Maraba to interview James "Jimmy" de Senna Simpson, the treasurer of the State Rural Producers' Union, and Diogo Naves Sobrinho, president of the cattle association. Tired of being vilified in the world press, attacked by Greenpeace, and besieged by the landless movement trying to occupy their lands, neither welcomed us warmly.

"I obey the law. I make money," said Simpson brusquely, by way of introduction.

There was no reason to doubt either statement. But these facts underlying his cattle business spell doom for the trees of the Amazon, unless a way can be found to reconcile the rights of these law-abiding citizens, participating in a free-market economy, with the perceived environmental needs of a larger group in the region and those far removed from the site. To the extent that Amazon deforestation affects global warming, these lawful activities are the culprits. Under what theory do you punish someone for obeying the law?

Simpson, whose family emigrated from Scotland in 1948 to work on cattle ranches in the south, arrived in Pará in 1992. About fifty and balding, dressed in a polo shirt and neatly pressed jeans, he sat behind a clean desk in an air-conditioned office in the empty fairgrounds, speaking with confidence about his cause. "Amazonia isn't what they say it is outside of Brazil. It's not full of criminals or being destroyed. It's growing and developing. It's the last agricultural frontier, whether that's what the world wants or not."

There is a seemingly endless supply of land in the Amazon and a seemingly endless demand for meat in the world. Between 1997 and 2003, the volume of beef exports from Brazil increased from about 232,000 to about 1.2 million metric tons. The increased supply has led to a steadying, a reduction even, in beef prices in Brazil and the export market, making this source of protein more affordable and, thus, more available—perhaps a very good result. Additionally, these exports have provided approximately $1.5 billion of sorely needed foreign exchange earnings for the country. Simpson reminded us that Brazil had become the largest exporter of beef in the world. "And this is why the United States doesn't want the Amazon to develop. Competition."

For Simpson and others, the world's environmental concerns are disguised attempts by the economic interests in the developed world to impede the growth of the Brazilian economy. Simpson sees an innate hypocrisy among the environment's advocates. "Who buys our wood?" Simpson asked. "You do. Europe does. Europe is the greatest destroyer of the forest today. Don't buy our wood, and we won't have incentive to cut [it]."

These arguments once were defensive posturing by rapacious ranchers in the Amazon. Most observers thought that the tension would eventually evaporate. Brazil would tire of propping up business that was not only unprofitable but also internationally unpopular. The ranchers would return to their homes in the south. Didn't happen. Nature bent to the will to succeed—and to profit. The ranchers who stayed, through years of trial and error with different types of pasture grass and with different cattle breeds, now present a more perplexing dilemma. When economic activity, which alters the environment, becomes profitable, the issues are joined. In the 1970s no one other than

the generals could justify the massive deforestation caused by ranching. Today, the free market justifies it.

In our interview with Diogo Naves Sobrinho, the cattle association
president said, "There were less than one million head of cattle when
my family settled in Pará in 1974. Today there are more than sixteen
million. There are nineteen slaughterhouses, and each one employs
many people. Twenty years ago we had 1.5 head per hectare, now it's
like five, even seven. What does that mean? It means more production
on less land."

It also means more land in use. The process starts with access,
and, unsurprisingly, the Amazon's road networks doubled from 1970
to 2000 by over fifty thousand miles. Eighty-five percent of deforestation takes place within thirty miles of a road.

There's no end to the ripple effect caused by the profitability of cattle ranching. If the end use is profitable, the vertical buildup also is
profitable. It's not unlike the cycle of cocaine, with land as the raw
product refined into cattle ranching. Everyone in the cycle who clears
the land, consolidates small holdings, or grows on it will make a profit.
Land grabbing, or *grilagem,* has become a way of life in the Amazon
and involves armies of hired gunmen who evict settlers on remote
plots of land, or the falsification of titles, or the bribery of land officials
to change recorded titles. Even this illegal process eventually becomes
legitimized, as Simpson's declaration suggested, by downstream bona
fide purchasers, unaware of any title defects. Challenging the claims of
people like Simpson through title searches in this morass of record
keeping is impossible.

The push into more remote areas of the Amazon, where land is
cheaper and ultimate profits greater, takes land acquisition away from
government controls. Statutory restrictions on the percentage of forest
that can be cut are meaningless in an area that government officials
cannot reach without days of travel. And, even if a government official
traveled days to a remote site, he would have to confront armed loggers all too ready to threaten him with violence or, more benignly,
ready to offer a bribe. The government official faces a dilemma that

would trouble any reasonable person: assert authority to save an unidentifiable tree on unmarked property, or protect his own safety and welfare. The government lacks the capital and human resources for these enforcement efforts. If there's an idle policeman in the country, should he be sent on a four-day trip to stop people from cutting trees, or to a street corner in a Rio neighborhood where fifty murders occur each month? From January 1987 to December 2001, 3,937 children were shot dead just in Rio de Janeiro.

Allocation of resources ranks among the most nettlesome problems the government faces, especially in this democracy where people, not trees, vote. The allocation issue underlies every aspect of social well-being in the Amazon. If education money is available, should teachers be sent to a rural area where the population is sparse and where reading and writing skills are less pertinent to daily life than they are in an Amazon city? If the choice is in favor of an Amazonian city, then why not choose a larger one in the more populous south? The same equation underlies decisions on how to provide health services—assuming you can only afford one, should a clinic be established to service a crowded community or a sparsely populated expanse in the rain forest? Environmental enforcers, too, can probably affect more lives by intercepting sources of air and water pollution in crowded urban areas than by trekking through the jungle to confront nimble loggers. Money or the lack of it, the diplomat Everton Vargas explained in Brasília, dictates the vexing choices the government must make and thus the consequences with which it must contend.

Naves explained to us, "Look, the municipality of São Feliz is nine and a half Belgiums. How do you police that? People come in, invade the land, get title, then sell the land and move on. That's a business in itself. Who benefits? No one. But when the area is so large and it's not even mapped out, can the government do anything about it?"

The World Bank report points out that even if the government wanted to send in a diligent, brave enforcement official to try to stop the deforestation, it would be too late. "Deforestation is taking place undetected by remote sensing. During the first year of deforestation, smaller trees are felled and grass is planted simultaneously: a laborer distributes seed in areas while mechanical earth movers 'clean' the land. A year after the grass has taken root under the trees, cattle are introduced into the area. Thus, livestock takes over the forest without, as far as the state is concerned, the trees being felled. The grass is burned during the second year as part of a secondary 'cleaning' of the forest. Medium sized trees are felled, leaving only the larger ones. The grass grows again (its roots having survived the burning process) and this enables once again the cattle to graze on the spoiled area. Only in the third year does burning take place which destroys what remains of the initial forest cover, thus permitting detection by satellite. Following this model of deforestation any action by the state is incapable of reversing the destruction that has already taken place or preventing the remaining forest from being ruined." Those who sit distant from Brazil and criticize its patrimony over the rain forest might perceive from this cycle of deforestation how complex the challenge of conservation is. Even though enlightened laws have been put on the books in the last twenty-five years and enlightened, dedicated people have come into the government and the NGOs, the trees fall at an alarming rate because holding on to them can be like hugging water.

The labor situation also mars these early stages of cattle ranching. Brazil abolished slavery in 1888, but between 1995 and 2005, more than five thousand slaves were liberated, most of them in the Amazon. In 2003, Binka Le Breton, a journalist who has written several books about the underclass in Pará, published *Trapped: Modern-day Slavery in the Brazilian Amazon,* which describes how workers are lured to work on remote ranches and are then forbidden to leave. The workers incur debts to the land grabbers or small ranchers, who provide them with food and shelter. There's precedence to this system. The Brazilian writer Euclydes da Cunha described the rubber tapper system as one where "man . . . labors to enslave himself." The Catholic Church estimates that there are still twenty-five thousand slaves in Brazil, ten thousand of them in the state of Pará alone.

The issue of slavery irritates Simpson. "This story about slavery doesn't exist," he said. "No one works when he doesn't want to. Why do the media publish these stories? Well, just to make media. Most of these cases aren't even real. The media call it slave labor when the workers leave without finishing. Of course they won't get paid until all the work is done! That's natural. Would you pay the construction workers building a new porch on your house before it's done? Of course not." Simpson's denials are forceful but not convincing. What happens beyond the tree line is anybody's guess.

The self-assuredness of Simpson and Naves, shared by the ranks of successful ranchers and farmers in the region, is a product of their experience. In some ways, who can blame them? Like their counterparts in Sinop, they traded a familiar setting for the jungle. Against all odds, they succeeded. No wonder they have no patience for those who sit and criticize.

They use the Internet and e-mail to share information about how to reclaim degraded land and how to select grass that gives better pest protection, that prevents invasive vegetation, and that allows better pasture management. The ranchers market their products together in order to attract slaughterhouse development nearby, thus reducing transportation costs and providing an anchor for job growth and economic development. As the World Bank observed in its report, "The experience of the American West would appear to hold lessons for what is occurring in Amazonia—initial economic failure does not impede expansion of the 'frontier' but rather speeds up the process of adapting to new managerial and technical methods."

Much of the science responsible for the enhanced profitability of cattle ranching has come from the opposite end of the Amazon—from the state of Acre, where there are three head of cattle for each person. Judson Valentim, who runs the EMBRAPA office outside the capital city in Rio Branco, has been as responsible as anyone for the research behind the methods that moved cattle ranching from red to black ink. EMBRAPA is a quasi-governmental cattle and agricultural research organization that partners publicly funded research with private enterprise. Valentim has been an itinerant throughout the Amazon, seeking out problems common to the ranchers, then producing a steady stream of articles on topics such as grass species, pasture management, pest control, and breeding techniques.

His work with grass, for instance, shows how this research easily translates into sound economic planning. As Meggers, Goodland, and other early critics of development argued, persistent Amazon rains quickly wash away nitrogen. The nutrients in many areas of the region occupy a thin layer of fertility on top of badly leached subsoil, which is a composition ripe for slash and burn agriculture. When Valentim, the son of a rancher from the state of Minas Gerais, arrived in Acre in 1979, he observed that productive pasture only lasted about five to six years; after that, the soil became infertile. Experimenting with tropical grasses (*leguminosas*), such as kudzu and a type of alfalfa, he prolonged the productive life of pastures to twenty-two years without the need for the repeated burnings that would eradicate invading plants and briefly replenish the soil.

Valentim advocated against burning as a means of pasture clearing, a radical departure. He showed that burning often fails to kill the roots of the burning plants, kills off the remaining productive grass, and acts as a trigger for the dissemination of seeds of invasive plants through the smoke. Burning also destroys the protective shield around waterways, leading to uncontrollable erosion and flooding. Because fire burns the flora indiscriminately, it robs the pastures of opportunities for shade, a necessity for cattle in this area. Valentim's recent study shows the correlation between shade and increased grass cover in the dry season.

He and his colleagues at EMBRAPA have made breakthroughs in other areas, such as artificial insemination to produce more productive herds, and developing patterns of pasture rotation, similar to crop rotation, to avoid overuse and depletion. These improvements have transformed the cattle economy in Amazonia and may soften its impact on the environment. With the ability to plan for maintaining a pasture for twenty-two years rather than for only five or six, a rancher can redirect efforts to replenish his land inventory and make additional investments on his land, knowing that his time for repayment has quadrupled, reducing the pressure to expand subsidies and ostensibly slowing the rate of changeover from forest to pasture. The more durable grass species have also increased the productivity of the land on which they are planted, leading to a higher head-per-hectare yield. "We can raise 2.5 to 3 animals per hectare," Valentim told us. "That's better than New Zealand." He added, "We can bring a steer to weight in half the time as the U.S."

The success of cattle ranching feeds on itself. With higher yields in the pastures and more stability in the area, attendant industries can be certain of uninterrupted supply. Slaughterhouses are more likely to locate in the Amazon, a far cry from the early days when cows would have to be shipped to the south, a journey that killed many of them. Unlike the ranches themselves, the slaughterhouses are labor intensive, which means improved employment opportunities will arise and benefit the area.

While cattle ranching has been the bane of environmentalism (and will continue to be), Valentim has successfully pushed ideas that are "good for the cows and good for the environment." His advocacy against burning is a good example, since fire has long served as the cheapest, most effective means of clearing. Techniques more advanced than fire may be unavailable to the small producer for years, but at least an alternative has been found. In addition, Valentim has ideas for the reforestation and restoration of degraded pastures, with a resulting increase in cows per hectare; these ideas provide alternatives to deforestation not previously seen.

Valentim expresses hope that environmentalists will reconcile themselves to cattle ranching and support his efforts to impose techniques on the ranches that make sense economically and environmentally. "Cattle ranching will not disappear," he said. "It is part of the tradition in the region, the culture." He noted that even on the renowned Chico Mendes Extractivist Reserve, families are turning to cattle in lieu of sustainable forest management. He joked about the heresy of this activity on sacred ground, but pointed out the obvious: raising cows is better than going hungry. He recounted a meeting he attended where rubber tappers were trying to persuade an Italian NGO to add cattle ranching as an industry it would finance. "One fellow got up," Valentim said, "and argued, 'Look, it takes a lot of work to extract products from the forest. I have to clear the forest, care for the plants, harvest them during the rainy season, and then I have to walk twenty-four kilometers [fifteen miles] to town, carrying them, in order to sell whatever I can. But, when I want to sell a cow, all I have to do is tell whoever's headed into town that I have one to sell. Someone will come to my door within a week with the price I ask and transport it to the city. The cow can walk, no one has to carry it.' "

This penchant for cattle raising at the grassroots level is cultural and blind to its impact on the environment. In rural areas, cows are like blue chip stocks: secure, less risky, and potentially more remunerative than alternatives. Valentim's account of his meeting included another anecdote about the connection between cows and a sense of security. " 'Look at my wife's leg,' one of the tappers said. 'When my wife got sick, I took her into town to find out what the matter was. The doctor there said it would have to be cut off. They couldn't do anything else. I sold a cow and took her to Rio Branco to see what they could do there. Same thing. I ended up selling a bunch of cows and went to Goiânia [a large city near Brasília], where they were able to help my wife. And look, she's still got her leg.' "

Just outside the dusty edge of Rio Branco along BR-364, the landscape turns an intense, luminous green. Not from trees, which are far in the distance when visible at all, but from the pastures that cover an endless line of rolling hills. This is cattle country. Clumped under shade trees or meandering along the ridgelines, hundreds of big-humped white Nelore steers forage on lush saw grass and swat their tails. The Nelore, the most common Amazon cow, originated in India. It is white and has oversize ears and a small hump behind its neck. The ranches run far back on either side of the new blacktop road that extends for 150 miles all the way to the border with Peru. There are hundreds of ranches, portioned off by fragile fences that hold nearly two million head of cattle in Acre.

The landscape is pretty, with low hills centered under a blue sky puffed with clouds. It has the vastness of the American West combined with the lush, palm-fronded groves of cattle country in central Florida. This landscape is not the moonscape of our bus ride from Maraba to Belém.

Among the sprawling properties is the ranch of Asseuro Veronez, a five-thousand-hectare spread with four thousand head of cattle just outside Xapuri. His ranch looks like a scene out of *Dallas*—long white fences marking off lush pastures. Convinced the Amazon is the best place in the world to raise cattle, Veronez is critical of the Brazilian government for failing to conceive a coherent policy for land management, which would allow the industry to grow even more rapidly. "Cattle is a force that's loose, and attempts to hold it back have failed," said Veronez. "It is vital to the economy of Brazil."

From a São Paulo ranching family, Veronez, a veterinarian by training, came to Acre thirty years ago as a banker, making loans for the Bank of Brazil to ranchers on the emerging frontier. Unable to resist the lure of the land, he and his wife, also from a family of ranchers, went into the business. He also started up the first slaughterhouse in the state. The Rio Branco phone book now lists thirty.

Like his brethren in Pará, Veronez insists that none of his success came from government assistance. "We get no government subsidies, although we do benefit from the technology and research of EM-BRAPA, which is a service a government should provide," he said. "We do this because it's a good business, and because we're good at it."

He scoffed at the "green industries" supported by the state government, such as extractive activities or on-site value-added efforts such as the furniture factory in Xapuri. "Too much going to too few," he said. "Most of the public policy is diverted toward people of the forest. But the attempts to implement the forest-based projects haven't worked up to this point, with a few exceptions. So, people with land will continue to do what gives them the best economic return. I don't see how the extractive reserves will come to anything if they don't produce a profit."

What bothers Veronez is that these sustainable development schemes distract from productive work. The state and federal governments, he said, don't know how to manage the Amazon, so neither production nor preservation is being served. The government's involvement also exacerbates violence, as the landless come to expect something the government can't deliver. Disappointed, the landless are bound to lash out against the government or the landowners themselves. He said so far these conflicts have been confined to the eastern Amazon, but he feared they would soon arrive in Acre.

His primary concern in 2003 was the restrictions on how he could use his land. The federal government requires that he keep a percentage of his land in its intact state. The national forest law requires that 80 percent of primary forests be maintained in an intact state, 35 percent of *cerrado* and 50 percent of transitional (between forest and *cerrado*) vegetation zones. These requirements have been on the books for decades, but have been honored most often in the breach through well-worn loopholes. Still, they grate on Veronez.

"The eighty percent minimum is absurd and only causes more destruction because legal ranchers buy more land to cut twenty percent of each, and the illegal uses are undeterred," Veronez said. He suggested that the government show more creativity, such as by establishing a system of forest reserves where rights to preserve acreage could be traded, or deforestation rights could be purchased by the government.

Veronez's recommendations, whether one agrees with them or not, present a new level of sophistication to planning in the Amazon. Trade-offs of development rights, air rights, watershed set-asides— these are land planning concepts that recognize the inevitability of development. It is no longer possible to simply shout no, lobby the government with international pressure, and expect that cattle subsidies will end and the industry will die. The views of environmentalists such as Dan Neptstad, acknowledging the inevitability of occupation, and those of developers such as Valentim, understanding that an economy and an ecology must coexist, are intersecting; a dialogue, if not collaboration, is beginning to emerge.

"We want to be partners with the government. We don't want to be fighting. But laws that are not realistic will fail," Veronez added.

CHAPTER 14

THE BREADBASKET OF TOMORROW

—

Blairo Maggi's realm is roughly the size of Texas. He doesn't own it all, but he certainly controls it. Not only is he the governor of the state of Mato Grosso, he's also the state's largest landholder, its richest businessman, and the patriarch of its farmers, and he is also the greatest single influence on the economic development of all of Amazonia. His impact dwarfs that of Ford and Ludwig, who came here to conquer nature; his vision extends beyond his own lifetime and his own projects. No Brazilian, even a head of government, has ever approached the level of influence he holds over the Amazon. It's not hyperbole to say a good part the future of the Amazon already has been and will continue to be determined by the will of this one man.

His legacy is simple: he learned how to grow soybeans in a place where no one thought it possible, and he figured out how to ship them out of an inaccessible location to the world's consumers. In the process, he started a revolution that is transforming the entire jungle—for better or worse, it remains to be seen.

—

Rondonôpolis, where Maggi agreed to meet us, sits on BR-364, about three hours by car southeast of the state capital of Cuiabá. Rondonôpolis is named for an underappreciated Brazilian hero, Marshal Candido Mariana da Silva Rondon, who guided the Roosevelt expedition in 1914. Rondon spent forty years "in the Brazilian wilderness, mapping backlands, stringing telephone wires, and making contact with Indian tribes," according to Mac Margolis in *The Last New World.* "No one— not Stanley or Livingstone nor even Humboldt or Wallace—spent more time exploring the tropics. He navigated through nearly as much of the jungle behind Rio and São Paulo as Columbus did the seas." More than any other person prior to Brasília's creation, Rondon connected the Amazon to the rest of Brazil.

The road to Rondonôpolis provides many reminders of the transformation of the Amazon in a single generation, and the road highlights how difficult it is to talk about the Amazon in generalities. Leaving Cuiabá, the road runs past a gritty urban sprawl of auto-parts stores, truck stops, and clusters of shacklike residences scattered among these commercial establishments. Zoning laws evidently don't exist, or aren't minded. From time to time, there's the expected view of the Amazon, thick vegetation and tangled trees, although the state of the rain forest visible from BR-364 is nothing close to its primary state and a glimpse is so infrequent that we cynically thought of the views as occasional Potemkin stands of trees to promote tourism. As on the ride through Acre, the Amazon again becomes a landscape resembling Africa—gentle rolling hills of green dotted with palm trees— then Texas, with endless pastures of scrubland hosting herds of cows meandering lazily. And finally Iowa, with sprawling emerald expanses of crops, primarily soybeans and corn, stretching to the horizon. Twenty-five years ago when we traveled this road, we could not have imagined such a scenic revolution. Blairo Maggi's father did, and he and his son made it a reality.

The Rondonôpolis headquarters of Grupo Maggi sits on the Amazon's periphery on the fertile savannah land that rolls up from the southern farm belt of Brazil and gradually turns into thick forest (which is the translation of "*mato grosso*"). The savannah, once covered with trees, now sprawls halfway up the state of Mato Grosso and is the heart of the state's agricultural and cattle industry. Farm equipment stores and pickup trucks driven by cowboys in straw Stetsons dot the streets of the town. A sign advertises the opening of the county fair, complete with prize bulls and the latest in satellite-guided combines.

We had been warned that the Maggi executives might be combative with visitors, defensive about the common accusations that they were largely responsible for the deforestation of the Amazon. This warning was quickly confirmed. While we waited for Maggi, a man in his mid-forties approached us, wearing glasses, pressed work shirt and jeans, and cowboy boots. He introduced himself as Adilton Sachet, Maggi's boyhood friend and business partner. (He subsequently was elected mayor of Rondonôpolis.)

He asked why we wanted to speak to the governor. We showed him a copy of the Brazilian edition of our previous book, *Amazonia*, in Portuguese, to which the publisher had added the subtitle, *Shout of Alarm*, and he winced. "Is this why you have come here?" he asked. "To shout an alarm? An alarm about what?" We explained that we weren't shouting about anything, that we had just come to find out how the Amazon had changed since our last visit.

"You left here when we arrived," he explained. "This land was nothing when we came. You saw that." Sachet, like Maggi, is the son of immigrant farmers from southern Brazil who headed north in the late 1970s in search of expanses of cheap land. At the time, the rule of thumb was that farmers in the south could buy fifteen hectares in Mato Grosso for every hectare they sold in their home states. "Our parents saw a lot of land, a lot of sun, and a lot of rain. It was just a matter of time before we became smart enough to understand how to make them all work together, under control. We have the sun all the time, except when we have rain. And we always have heat. There's no place in the world that has growing conditions like we do."

We told him that there are a lot of people who don't like what he's doing, good growing conditions or not. They'd prefer forest over soy. He scoffed, "What do you know about the Amazon in Washington? Or Paris? Don't you realize that we are the greatest preservationists of the Amazon? If we kill it, then we kill ourselves. We have our families here. Just look at this building. We have enormous investments. You came on the road from Cuiabá. You saw the farms. You can see the investments we have. And it's not just money. It's human capital. We are not going to destroy what we built. We are going to improve upon it and pass it on to our kids."

We walked through the modern office building to the area in back, with a swimming pool, a soccer field, and volleyball courts. An evangelical minister was sermonizing on the virtues of hard work and persistence to an assemblage of employees and their families gathered for the inauguration of Grupo Maggi's office complex and recreation center. As the pastor went on a little too long about the confluence of God and the state of Mato Grosso and the Maggi family, some restless listeners slipped away to play soccer or bat a volleyball across the sandy court. They were politely chastised by a supervisor who shooed them back into the audience.

There is a mystique about working for Grupo Maggi in the Amazon. The employees see themselves as pioneers in a pioneering industry, visibly changing the world they live in. From a compound of airy and air-conditioned office buildings swept clean on the hour by an army of custodians, the employees of Grupo Maggi communicate by cell phone and Internet with managers of Amazon ports, barge captains on the river, grain dealers at the Chicago Board of Trade, and shipping companies in Rotterdam and Shanghai.

Sachet left, and minutes later returned to tell us that Governor Maggi was indisposed, a bout of food poisoning from a political dinner the previous evening. We asked him if we should reschedule the interview. "Don't worry," Sachet assured us. "He'll be out. He will only allow himself a little time to be sick."

Finally, the door to the men's room opened and the governor walked toward us, a short dark-haired man in his mid-forties wearing white pants and a baby blue knit pullover soiled by sweat spots. We could see his discomfit; he crouched in pain, pasty white and sweating. His wife whispered to him as he walked, "You must go home, Blairo." But he was a politician, and he didn't want to be accused of ducking an interview.

"I am very sick," he said, "so I just want to make some statements and answer some questions. We can talk more another time."

He rapidly provided a synopsis of his philosophy. "I am not interested in critics. I'm interested in results. Remember what we are doing here," he said. "We are growing soy, which is the single most important source of protein in the world. We are creating the greatest soy growing area of the world. This is the next great breadbasket. Right here in Amazonia. We are growing crops that are basic to human needs—soy, corn, cotton, and rice. Without this expansion, the world will die of hunger. We are feeding the world, and we are doing it without destroying the environment."

The international media portrays him differently. He's become a symbol for the changing landscape of the Amazon. In May 2005, the English newspaper *The Independent* featured Maggi in a story titled "The Rape of the Rainforest . . . and the Man Behind It." Maggi responded with a press conference in which he declared, "I am not a rapist."

In the months after that first meeting, we visited Maggi several times, having been warned after making each appointment, "the governor is always on time," a decidedly non-Brazilian trait. Even in the face of criticism from the world's press, he never lets go of the messianic goal of his industry. He posited the question, "Do you want trees and hunger? Is that what you want? Because if you leave the forest alone, that is what you will get."

Pitching the alternatives in absolutes makes for effective political sloganeering, but, as Maggi himself revealed, the trade-offs have become much more nuanced. Arguments about Amazon land use traditionally have proceeded on parallel courses: you're either for development or you're for the environment. When deforestation rates stood at around 3 percent twenty-five years ago, the outcome of the arguments really was academic, as development wasn't going to dent the forest. Environmentalists, bolstered by Meggers's historical perspective, took a "do not touch" approach. Yet it proved to be an unrealistic and unsuccessful strategy. The forest shrank, and development techniques grew sophisticated. A "Save the Amazon" campaign clearly has to recognize the permanence of Maggi's revolution and the benefits it provides, or this campaign too will fail.

"People who don't live here don't understand what we are doing here," he told us in the comfort of the governor's office in Cuiabá. "I have been here for twenty-three years doing research on this land. While our parents were buying land and clearing land, we were learning how to use the land."

When we suggested that succeeding in this climate wasn't difficult, he practically snorted. "If that were true, then there would be no difference between here and parts of Africa. But we have succeeded and Africa hasn't. Why? Our people. We have studied what needs to be done to create things, not destroy things. We know that you cannot make an omelette without breaking some eggs, but we have made an omelet. Africa just has a lot of broken eggs."

Maggi's ambition stems from his wish to fulfill the vision of his father, who died in 2002. Andre Maggi arrived in Mato Grosso in 1979 after attempting to establish a foothold in western Pará. Looking for large expanses of cheap land, no longer available in the south, he bought one thousand hectares of cleared land from a rice farmer unable to pay his mortgage to the Bank of Brazil. The Maggi empire began.

The qualitative difference between Andre Maggi and the landowners we encountered in 1979 was that most of them were absentee owners, uninterested in establishing stable communities or sustaining economic activities, since government funds were readily available to cover losses. Maggi's father and his compatriots moved their families to the Amazon and literally bet their lives on their success in the new land. They didn't have public money to squander or nice homes to return to.

At first, Maggi's father pursued rice as his crop of choice. Although lacking soy's economic potential, rice was adaptable to this soil, and banks, familiar with the crop, were willing to finance storage facilities and equipment purchases. Maggi's first soy planting failed because he used seeds from Paraná, which were not suitable to the aluminum-rich soil of Mato Grosso. Convinced, however, that only the right seed was missing from an otherwise perfect setting for large-scale crop cultivation, he experimented incessantly with different varieties of seed and attempts to alter the composition of the soil. The high aluminum content of the soil, he figured, could be overcome with a combination of additives such as nitrogen, phosphorus, and lime, and with crop rotation between rice and soy. After four years of trial and study, he saw results—and it was then that he showed the true business genius that would make Grupo Maggi such a formidable enterprise. Realizing that his own success would lure imitators, he amassed capacity at every level of production. He formed a seed business and a storage business, and he built a hydroelectric plant to provide farmers with power at a reasonable cost—vertical integration in the heart of the Amazon.

Andre Maggi worked the land himself, preferring to sleep in the laborers' dormitories rather than in his house in Rondonôpolis. Besides experimenting with the soil, he also kept an eye out for acquisitions, as he appreciated that as the size of farms increased, the marginal cost of growing soy decreased. He found a ready market. Many farmers failed in their attempts to adapt to this new soil or couldn't afford to expand their holdings to a size that made economic sense. Governor Maggi insists that his father never tore down a tree; he just bought cleared farms that had failed or whose owners had decided to move on.

The son took the enterprise to a new level. He developed networks with other producers so they could share knowledge of the best seed varieties. He brought agronomists from around the world to seminars on tropical agriculture. And he promoted the farmers' successes to attract the necessary infrastructure for large-scale agriculture—the machinery vendors, the banks, the multinational commodities companies looking for new product sources. A service industry developed. Concentric rings of economic activity emanated from the soybean. Blairo Maggi became the visionary among the agribusinessmen. He promoted cooperation among them for activities too expensive for individuals, such as research, transportation, and storage facilities. He anticipated their needs for growth and implemented action plans that provided easier access to equipment, financing, and, ultimately, markets for their products. These action plans were primarily funded with his money or with a consortium of financing he assembled for their common good. The soy, meanwhile, kept growing.

Brazil's sudden emergence as a soy superpower stunned the world. In 2003 the United States Department of Agriculture issued a report concluding, "It is apparent that the underlying potential for growth in Brazil's agricultural economy over the next century has been seriously underrated to date, given the nation's wealth of available land resources and its highly professional and enterprising agribusiness community." The president of Nebraska's farm bureau described Brazil's potential as a "time bomb" readily able to meet the world's soy demand much more competitively than his own constituency. *ProFarmer* magazine printed the notes of a farmer in Bremer County, Iowa, in February 2001, on a visit to Mato Grosso: "I went down to Brazil with some preconceived notions of how they farmed. I came back a different person. We are not prepared as farmers. There are people in South America who will change the way we live up here."

In the 1990s, annual global soy consumption grew by about 5.5 percent per year. At the same time, Brazil's government fostered domestic production by reducing or eliminating import barriers on agricultural products such as fertilizer, pesticides, and machinery. It also removed export taxes on soybeans and soy products. World production in 2005 totaled 219.7 million tons. Brazil produced 62 million tons, or 28 percent of the world's total, second behind the United States with 78 million tons. By most estimates, Brazil will surpass the United States in production by 2010. In 2004, Brazil earned ten billion dollars from soybean exports (down slightly in 2005 due to crop damage).

During this time, mad cow disease appeared in Europe. Soy became the exclusive diet of livestock. Also, China's rapidly growing middle class started to show its appetite for beef, and from 1999 to 2003, China's imports of soy grew tenfold from 620,000 to 6.1 million tons. Only Brazil, and primarily the Amazon, could meet this increased demand quickly. These two surges in demand occurred when the American dollar was strong against Brazil's real, giving Brazil's farmers yet another competitive advantage.

Brazil has also benefitted from Europe's antipathy toward genetically modified soy, which constitutes much of the U.S. production. In 2004, 63 percent of European soy imports came from Brazil. It remains to be seen how Brazil's changeover in 2005 to allow for the production of genetically modified soy will affect this commerce.

The corner has been turned. The U.S. soybean farmers will no longer be able to compete with the Amazon. In 2004, the U.S. Department of Agriculture reported, "The combination of low land prices and high crop yields gives [Amazon] farmers a healthy average gross margin estimated at 15–30%, despite some of the country's highest transportation costs." North Dakota recently commissioned a study to determine its ability to compete with the Mato Grosso farmers, and it concluded, "Soybean production in 2003 was projected to be over three times more profitable per acre for Mato Grosso than projected for North Dakota. For Iowa, soybeans show a potential substantial loss for this budget that reflects all economic costs of production." In 2003, soy farmers in Mato Grosso returned $51.97 per acre to management. In North Dakota, farmers returned only $15.97, and they lost $44.97 per acre in Iowa due to the higher land costs.

The most startling conclusion of the North Dakota study addresses the sale of a bushel of soybeans in Rotterdam. American farmers have always enjoyed a vaunted advantage in transportation costs. No longer. In 2003, the total cost in Rotterdam of a bushel of soybeans from North Dakota was $5.76 (freight accounting for $1.17), from Iowa $7.21 (freight accounting for $.93), and from Mato Grosso $4.57 (freight accounting for $1.33).

No region in the world compares to the Amazon for its continual supply of heat, sun, and rainfall. And no other region has such an endless supply of usable land. Some areas of the Amazon see three crop yields per year. In the winter, satellite images show a white and frozen American Midwest lying idle, while the Amazon blossoms. This competitive advantage will only grow as farmers introduce newer and more efficient soybean, corn, and cotton varieties that will lead to increased yields. The 2003 soybean variety (BRS-Raimunda) yields an average of five tons, doubling the production of existing varieties. Genetically modified soy will increase yields even more quickly and decrease costs through lower herbicide use. And it's not only soy: new corn varieties will allow farmers to yield fifteen tons per hectare, compared to the existing average of nine tons. And from 1997 to 2003, Mato Grosso jumped from producing 6 percent of Brazil's cotton to producing more than 50 percent.

Who would have thought the American Midwest would lose out to the Amazon?

In May 2005, *The New York Times* published an editorial called "The Amazon at Risk," which argued: "Right now, for instance, the biggest single threat to the Amazon is the explosive growth of soybean farming in the state of Mato Grosso on the forest's southern fringe, fueled mainly by soaring demand in China and Europe. As it happens, Mato

Grosso's governor, Blairo Maggi, is also its soybean king—*o rei da soja*—who has been quoted as saying that a 40 percent increase in deforestation in Mato Gross 'doesn't mean anything at all, and I don't feel the slightest guilt over what we are doing here.' " The *Times* editorial was in response to Brazil's deforestation figures from 2004, showing that half of the sixteen thousand square miles of deforestation occurred in Mato Grosso. The state accounted for half of the deforestation in 2003, too.

But Maggi keeps a different set of statistics. The per capita income in Mato Grosso has risen more than fifteenfold in the last twenty years. The productivity of soy has grown in Mato Grosso from 1.57 tons per hectare in 1980 to 3.1 tons in 2003. In 1980, 56,000 hectares of soy were planted, compared to 4,521,000 in 2003. Maggi's stated goal: 40 million hectares under cultivation.

Maggi often refers to an argument that appeared in a different *New York Times* article. In September 2003, the newspaper reported that the "world's wealthiest nations give more than $300 billion of subsidies every year" to their own farmers, more than the gross national product of sub-Saharan Africa. Subsidies, Maggi argues, are more venal than deforestation. The *Times* went on, "These payments allow industrial-size farms to produce many more acres of crops than are needed for domestic consumption, and they are sold overseas at low, subsidized prices. Farmers in developing countries cannot compete with cheap imports. They lose out in their own markets and have little chance of exporting."

It is not difficult, sitting with Maggi, to understand how people living in this place see the international environmental movement and the outcry against deforestation as a plot to send these farmers back to a slash and burn economy.

"What are we doing here? We are feeding the world. We ask that the developed countries help us by financing our infrastructure, but they cry about the environment without understanding the environment. We ask them to help us by dropping the subsidies they give their farmers, and make it easier for us to compete, so perhaps we wouldn't have to expand so much. But they don't do that. They are maintaining the levels of poverty in the world by making it difficult for us to realize our potential. There is a hypocrisy in Europe and in the United States. They cry when we cut a tree, but they don't cry when children die or do not have an education. If they want to help us, then help us help ourselves," he argues.

His resentment is palpable. A farmer in Iowa would never tolerate being told not to touch 80 percent of his land or that no one will reimburse him for leaving it alone. Imagine German, Dutch, and Japanese scientists regularly visiting that Iowan's farm to study when his cows defecate and how that affects the land. And when and if they see an adverse effect, they alert the world press. Imagine that when the Iowa farmer wants to sell his products in Germany, Holland, or Japan, he has to compete with domestic products heavily subsidized by those governments. Or if he tries to sell his crops to a third country, he then has to compete with these same farmers whose government subsidies make it impossible for him to make a profit. How long will an Iowa farmer tolerate this interference with his land use? Or these artificial barriers to entry in these markets?

To Maggi, the environmental movement represents the Trojan horse filled with the vested agricultural interests of the developed world. These countries have already destroyed their own environment, Maggi argues, so he is cynical about their stated concerns. He believes these so-called environmental concerns espoused by the developed countries serve as surrogate arguments to protect national commercial interests.

Thirty-two-year-old Ane Alencar is a Brazilian environmentalist who works with IPAM in Belém, an NGO that has sought Maggi's cooperation both in establishing monitoring stations on Grupo Maggi land and in preparing a master plan for the territory for the eventual paving of the Cuiabá-Santarém road.

Alencar symbolizes the by-product of Sarney's invitation to host the Rio Summit and his embrace of a Brazilian environmental agenda. The attention drawn to the Amazon by the events of the late 1980s and early 1990s led to environmental-program funding in the federal university system, which attracted a generation of public-policy-oriented students. Likewise, the end of the military government provided an opportunity for socially conscious students to channel their energies to areas other than political dissent. As a consequence, a well-educated, highly motivated cadre of Brazilian environmentalists has emerged. Their efforts remove the oft-heard criticism that the environmental movement is the product of foreign conspirators.

When we met her, Alencar had never met Maggi and confessed to us that she had mixed feelings about him. "He clearly understands, as a politician, that public opinion is important even in business. And that is good. An example of this is that the governor himself has done a good job in implementing a certification program for his soy." The certification program requires that producers not plant on illegally deforested land, agree to policies for working conditions, and use only certain chemicals in planting. In return, certain purchasers buy their product, often paying a small premium. It's similar to the pioneering "certified products" programs for hardwoods that Governor Viana described, where volume purchasers such as The Home Depot are encouraged to refuse to buy nonconforming products—an attempt at marrying consumerism with environmentalism.

Maggi agreed to a soy certification program, according to Alencar, but he does not require the producers he finances to follow the program, ostensibly because he doesn't want to alienate them. "We can make all the publicity we want that this is a good program," she told us. "That doesn't help. All he has to do is say, 'No certification, no financing.' " She added, "I don't know much about politics, but this is politics."

Maggi has also frustrated environmentalists by delaying the implementation of a "zoning system" for Mato Grosso, in which only designated areas could be deforested and planted. Others would be set aside and preserved. The plan differs from the 80-20 set-aside program in that it addresses tracts of land under multiple ownership and allows for large-scale collective land management rather than piece by piece governance. The latter governance has failed over the years, despite its superficial appeal, because restrictions on individual tracts can easily be avoided by selling off pieces of land, with each subsequent owner cutting another 20 percent of what was once protected.

The concept of regional land management butts up against Brazil's tradition of exalting individual landownership and monarchical prerogatives over the common interests of the community. It remains to be seen to what extent individual entrepreneurs are willing to sublimate their goals of short-term profits in order to benefit the greater good of the environment and, perhaps, in order to attain longer-term profits. The population of Amazon agribusinessmen, natural risk takers and self-made successes, does not have characteristics compatible with long-term planning. However, Maggi's and Sachet's references to this eventually being the home of their "children's children" may indicate that this ethos is changing.

"This plan will go a long way in controlling the patterns of deforestation in his state," Alencar said, "which is important in microclimate change, as well as in just keeping strategic areas in preservation. The plan is ready for his signature. I don't know why he won't sign it."

IPAM is at the forefront of this area-specific approach to preservation. In a separate interview, Dan Nepstad, one of IPAM's founders, said that the traditional piece by piece management was outdated. He said that a justification for the development of that regime, other than ease of enforcement (oversight over a single owner certainly is easier than coordinating the interests of multiple owners and imposing a land use plan), had been to maximize species preservation. He has concluded that that is an "overrated" motivation for land use management, citing as an example that 93 percent of the Brazilian Atlantic rain forest has been wiped out and only one species of bird has been lost. Nepstad, who acknowledges the "inevitability" of mechanized agriculture and large-scale cattle ranching, has focused IPAM on preserving forests where they would have the greatest impact on climate conditions in a specific region, on preserving watershed areas, and on reforesting degraded areas that have no alternative productive use.

"The governor also has the ability to make sure that the state reserves, whether they exist now or come in the zoning system, stay preserved," Alencar told us. "Maggi has to protect what should be protected."

This last point has provided perhaps the greatest embarrassment of Maggi's administration. When we met with him in 2003, he described the process by which he chose his state's environmental secretary. "I included the environmentalists in the process," he explained to us. "I asked the environmentalists who they wanted me to appoint as the environment secretary and who they didn't want me to appoint. I studied their lists, and then I chose my environmental secretary—the person they most *did not* want me to appoint." In June 2005, Maggi's environment secretary was arrested and charged with taking payoffs in return for allowing loggers to deforest state reserves. Maggi claimed he was "shocked" and fired him.

In the wake of the scandal involving his environmental secretary, Maggi pledged to support a three-year moratorium on deforestation in Mato Grosso. If it comes to pass, the rates of deforestation should plummet, although some worry that pent-up demand will spill over into neighboring states. Interestingly, a deceleration in deforestation rates should not affect soy production. The U.S. Department of Agriculture "has conservatively determined that tropical soybean production in the Amazon could expand by more than 40 million hectares on existing non-forested land by utilizing a portion of current pasture and savannah land resources." Production of soybeans in the Amazon could increase by as much as the size of the states of Illinois, Iowa, and Kansas combined without a single tree being torn down.

Maggi complained he has trouble getting this point across to the media. "What I am saying is that soy is a productive use of unproductive land," he told us. "This land already has been deforested and is not being used. We should use it. We don't have to take down any trees to expand the soy frontier."

Besides large-scale land reclamation, Maggi advocates what he calls a "revolutionary" idea of preservation, a "zone of exclusion." He studied the map of the Amazon on the table in front of him. "If you take this line," he said, drawing his finger along the Tapajos River, "and say that there's a moratorium in all the area west of the Tapajos region, then you won't have any problems. You touch nothing in this area. Untouchable. You will preserve the rain forest forever. And if forever is too long, then have the moratorium for fifty years and then come back and see where we are."

He emphasized his seriousness. "You want to save the rain forest? This is how to do it. Don't say it can't be done. Just do it. But it will never happen, because the Ministry of the Environment and their friends don't want a solution, because then they would have nothing to do."

Whether Maggi's proposal would dissolve the Ministry of the Environment misses the real conundrum embedded in the idea—that is, the tension between the need to preserve the environment and the need to provide opportunities for those already living there and for the disenfranchised settlers yet to come. The moratorium strikes Maggi's fancy because it is a simple, all-encompassing proposal, easy to espouse, superficially "green," and impossible to police. It has become his standard response to critics because it puts them on the defensive. It also serves his obvious self-interest; by slowing down the spread of the frontier, he would impede the growth of competing soy producers and maintain his dominance in the industry.

The truth is, Maggi expresses no guilt about what he has done. His father's recent death has caused him to reflect on his own life. "What have I done wrong?" he asked. "With my father, I have shown the world that this is the next breadbasket. I have made my family's life better. I have given jobs, good jobs, to many people. The other farmers have made their lives better and created jobs. The towns of my state have schools, hospitals, and not so much malaria and disease. We have fed people. We are fighting poverty in the world by making food cheaper and trying to give farmers in other countries an opportunity to compete."

People often miss the lasting impact Maggi will have on the region, despite the large number of hectares he farms (in 2003, he had ninety-five thousand hectares of soy, twenty-six thousand hectares of corn, and twenty-seven hundred hectares of cotton under cultivation—about five hundred square miles). Other farmers may one day farm more land, but it is Maggi who solved the single most significant impediment these farmers faced: access to markets. It was only after Maggi created a transportation network that the trees really began to fall. "We were strangling on our product," Maggi told us. "No more."

His vision, a four-part transportation strategy, is already in place, though major expansion plans are in the works. The first corridor consists of trucking the soy from the farms in Mato Grosso and lower Rondônia along BR-364 to the Madeira River city of Porto Velho and then on to a unique grain port on the banks of the Amazon River itself. Maggi's group controls the traffic into Porto Velho. The trucks rest at a truck stop until slots open for them at the storage facility along the port. The trucks are weighed at the point of loading and then weighed again at the port to make sure enterprising drivers have not lightened the load. Twenty trucks per hour, fifty thousand kilos per truck. The grain travels to four silos, each with a total storage capacity of forty-five thousand metric tons. The trucks then proceed to another garage and take on lime or phosphate. In Maggi's world of commerce, where unused capacity means lost profit, the trucks haul this fertilizer back to the soy farms, only to begin the journey again. Twenty-four hours a day, seven days per week.

Meanwhile, over on the banks of the Madeira River, Maggi's barges queue up at the loading area. Each has a draft of fifteen feet and a capacity of 130,000 bushels. The typical Mississippi barge has a draft of nine feet and carries about 50,000 bushels, a comparison not lost on Maggi. Maggi borrowed the design for the towboats from Finland, where ice floes, not the logs found in the Amazon system, present impediments to navigation. The locks on the Mississippi River, six hundred feet long, take about five to six hours to disassemble and reassemble. Delays along the Mississippi cost more than one hundred million dollars per year, and the lock system is over eighty years old. It would cost hundreds of millions of dollars to replace—after years of study and environmental impact statements. There are no locks on the Madeira River. Maggi built his entire system in three years at a cost of one hundred million dollars with almost no publicity—to avoid interference from environmental groups.

Maggi strings nine barges together, three rows of three, and places a tugboat behind them. Linked to a satellite, they travel without the need for the crew to do anything but watch. The river never freezes over. Some navigational perils appear in the dry season, but the pilots' skill, combined with the satellites, minimizes the potential problems. The convoys never stop. The grain travels up the Madeira, then makes a right turn where the river intersects with the Amazon, and lands seven hundred miles away at the port of Itacoatiara. Unload the grain, load up on fertilizer. Fifty hours from Porto Velho to Itacoatiara, 106 hours on the return. The crews work forty-five days on, eight days off.

At Itacoatiara, which is three hours by car on a decent road from Manaus, Maggi quietly built a floating port facility in 1997, which can handle oceangoing grain carriers twelve hundred miles upriver. The Amazon River at that point is seven miles wide and 130 feet deep, and because the river rises and falls 65 feet from rainy to dry season, the docks must float. Gigantic vacuum cleaners suck out the grain from the barges at the rate of nearly seventy-five thousand bushels an hour. At the same time, ships from around the world are loaded—three per month, each carrying away almost 60,000 metric tons of grain. In 2003, more than 1.3 million metric tons of soy left this port.

About a mile down the river, Maggi has a smaller port for receiving fertilizer, and once the barges drop their soy, they go here. Then they complete their trip to Porto Velho. While most of the fertilizer comes from Israel, when it arrives in Mato Grosso, it still has a net cost to the farmer of about half of what the same domestic product would cost had it been transported from the ports in the south of Brazil.

The most controversial branch of Maggi's transportation plan involves the paving of the eleven-hundred-mile Cuiabá-Santarém Highway, BR-163. Because 85 percent of all deforestation in the Amazon occurs within thirty miles of a road, BR-163 has generated enormous concern, particularly the feeder roads it will spawn into some of the best soy-growing areas in the Amazon. IPAM has predicted that paving BR-163 will deforest between 2.2 and 4.9 million hectares within twenty-five to thirty-five years.

Because of the importance of the road to the farmers and the controversy it has engendered, Maggi organized a consortium of local farmers to pay for the paving. Maggi committed about $300 million of his state's funds to complement $125 million of private funds, an unprecedented public-private partnership. The government funds were not taken from the general budget but were raised directly from the farmers through a tax on soybeans, cotton, cattle, and wood. The consortium will be allowed to charge a toll on some of these roads, but they expect their return to show up in the substantial reduction in freight costs and the appreciation of land values.

The other prongs on Maggi's map include a linkup to the railroad network of the iron mine at Carajas, so the crops can go to the modern Atlantic port of São Luís, and a rail link from the port of Santos in the south to Rondonôpolis and then to Cuiabá and Porto Velho. On a map, these routes make the state of Mato Grosso look like the epicenter of Brazil, with four broad rays shooting from its center. It's Maggi's version of the famous *New Yorker* cover showing a Manhattanite's view of the world.

Toby McGrath, an influential American environmentalist who works with Ane Alencar in Belém, sees Maggi as the catalyst for forcing a serious, unsentimental debate about the Amazon. Underlying McGrath's view of the Amazon lies a basic foundation that many twentieth-century scientists refused to acknowledge, hoping it would go away: the Amazon is occupied and will remain so. The Amazon's occupation is tied to an increasing population and to economic growth, the same factors behind the expansion of the American West. McGrath is critical of the preoccupation with the statistic that has alarmed most environmentalists: how much of the forest is gone. While the environmentalists "worry only about slowing the rate of deforestation, they forget to look at the quality of the lands that come after the devastation. They end up losing twice," he said. "First, they lose the forest. Next they lose the opportunity to take advantage of the natural capital of the woods to create a productive and sustainable regional economy."

McGrath, who once feared that the Amazon might succumb to the will of a politician from Mato Grosso, now worries less about the phenomena Maggi has created. "In the last fifteen years, government's capacity to control effectively what is happening on the land changed. The institutions are stronger, the environmental legislation is more effective, the political will is greater, and the monitoring technology is more sophisticated and diverse. But it is necessary to understand that this control must not be made to avoid occupation, but to guarantee that it occurs according to the legislation and maintains the regional ecological balance." McGrath agrees that Brazil doesn't need new laws, it needs enforcement of the laws it has, although he notes that Maggi has been lax at enforcing those laws.

Soy cultivation probably comes closest to approaching the sensitivities that McGrath advocates for any large-scale occupational activity in Amazonia. "It uses direct planning and occupies only flat areas, minimizing erosion. It doesn't use fire, like ranching and other agricultural activities. The social dynamic of soy is very different than the expansion of cattle ranching. The violence is much less common than in the case of loggers and land speculators. Also, the job quality on a soybean farm is in general much better than on an extensive cattle ranching farm."

McGrath gave an interview to *Veja*, Brazil's largest newsmagazine, in late 2003. He expressed these views, and found an unlikely receptive audience—Governor Maggi. Up until that point, Maggi had defended the increased soy production with the outdated argument, "We have enough land to continue to grow at this pace for three hundred years." The "enough land to grow" justification for turning the other cheek on deforestation was discredited a generation ago. Growth increased exponentially, rendering any present growth rates unreliable for predicting the future. More important, the quality of deforestation—where it occurs—has proven as significant as quantitative deforestation—how much occurs. After reading McGrath's arguments, Maggi ceased this line of reasoning and began to talk in terms of job creation and conservation management. Lines of communication were becoming visible.

Undoubtedly, Maggi controls the future of development in the Amazon more than anyone else. He controls the capital, and he controls his state's government. To write him off as a provincial oligarch in the third world would be foolish. To compare his vision to the monomania of Ford and Ludwig would also be mistaken. As Maggi himself said, those men had isolated projects that had no planning behind them, no experience. "I am involving thousands of people in what I do, and what we do is basic to the survival of the world. Their example is not relevant to what I am doing."

Maggi admits there may be limits to his growth. "There are two reasons we may not continue to succeed," he said. "First, the world may find other ways to get protein and fiber, so they won't need soy and cotton. It's possible—not very likely, but possible. That might be good for the world, but it would be very bad for us."

The other limit on growth is just as unlikely. "Someday they will develop varieties of seed that can grow in bad soil. China, which doesn't have as much arable land as we do, will be able to use these seeds and stop importing from us. But that will not happen soon, I don't think. In the meantime, we will grow what we have and the world will consume it."

As the largest individual supplier of protein in the world, Maggi's importance spills over the borders of Brazil. He has internationalized his operation in ways rarely seen in any developed country: growing soy in Brazil, shipping that soy to China, buying fertilizer from Israel, copying shiploading techniques from Switzerland, imitating barge designs from Finland, hedging his prices on the Chicago Board of Trade. No market in the world is too far away for Maggi. No idea in the world about improving his operation escapes his attention.

Maggi is a twenty-first-century Brazilian nationalist operating in a multilateral world without borders. He embodies a post–Cold War view of globalism: that business drives alliances and that political allies can be economic competitors. Brazil can befriend the United States or Europe without having to walk in their shadow. Yet the legacy of economic protectionism makes Maggi resentful. In a speech following September 11, 2001, Maggi articulated his view of the world's economic battleground as a mournful statesman surveying the carnage of that day. "The millions of poor and destitute in the world who live with less than two dollars a day find their ultimate refuge in religion," he said. "From there to fanaticism, it's only one more step."

Although he expressed his sympathy for the victims of the attacks, his message contained little pity and abundant warning. "[The] populations of the most powerful countries on earth will live under an empire of fear and threats from all orders. And the precautions being taken to avoid another occurrence will be for naught if the causes underlying the terror are not attacked directly. . . . This opening of markets in the name of globalization is not a two-way street. It serves the rich countries only by giving them access to markets that interest them and the means to protect their internal economies through the use of ingenious artifices." A bold pronouncement from the governor of a state most of the world's policy makers couldn't even find on a map.

In his accomplishments and in his ambitions, Maggi personifies that the Amazon story is more than a remote environmental squabble. The Amazon is about the future of Brazil, and Brazil's prominence on the world stage, long delayed, is inevitable. In the post–Cold War world, where a country's keen economic competition is more important than military might in proving that country's global influence, there is no reason to expect a lessening of nationalistic pride. The developing countries participating in a global economy advocate equity in setting the rules, and their collective clout is succeeding. In September 2004, Brazil won two landmark cases at the WTO, one claiming that the United States had paid illegal subsidies to its cotton farmers and the other that the E.U. had exported more sugar than it was allowed to under world trade rules.

Maggi won't countenance that European or American politicians, beholden to the lobbying efforts of his competitors, try to dictate his choices of land use and access to markets—through subsidies, pressure on the Brazilian government, or funding of international environmental organizations. If they don't want to buy his products, fine. He's confident that he'll figure out a way to compete against them. Just as he doesn't tell them how to run their businesses, he expects them to stay out of his.

CHAPTER 15

THOSE LEFT BEHIND

—

As much as hope and progress have come to inhabit the Amazon, they never have been abandoned by their constant companions—despair and poverty. As the forces of nature, human and ecological, clash throughout the region, some immigrants have emerged triumphant, but many more have been left behind. Searching for gold or even searching for the right cattle-grazing grass or soybean variety is never as difficult as the search for human dignity in the face of so much hardship.

Lost on the periphery of towns and barely surviving on the banks of rivers, a large population has been marginalized in Amazon towns, as in most of the rest of the country. They meander through a daily existence in unsteady houses, uncertain of their next meal and in precarious health. Many cannot read or write. They are like forgotten people everywhere, living in filth, preyed on by insects, suffocated by stenches, debilitated by hunger, and saddened by the wail of their suffering infants.

Over 70 percent of the population in the Amazon is urban, and typically the new arrivals ring the towns in miserable living conditions. They have difficulty adapting outside of the forest, surrounded by neighbors—also struggling—rather than by trees and wildlife. They generally lack formal education and any skills beyond manual labor. The urban poor are a phenomenon few expect to see when they envision the Amazon, although their numbers shouldn't be surprising. The ranks have swelled as a consequence of failed uses of the forest.

Altamira, a city of approximately one hundred thousand, is a place where the undercurrents that flow beneath the successes of the cattle ranches and soy farms have converged. The poor, displaced by the cycle of deforestation begun by loggers, have settled here. Located one hundred miles south of the main Amazon, about midway between Belém and Manaus, Altamira has no strategic importance, since the Xingu River isn't navigable to the north because of impassable rapids. The city grew out of a road stop where the TransAmazon Highway meets the Xingu, just a place where people stopped and ended up staying. Altamira acquired fame as a backdrop in the 1979 movie *Bye Bye Brazil,* which was made at a time when the military government was at its most repressive. In the movie, a wandering circus troupe, inspired by government propaganda aimed at would-be settlers, anticipates Altamira as a Shangri-la-like destination, a hidden tropical paradise. At that time, the Amazon was as foreign to most Brazilians as it was to Americans, so the joke held until the end of the movie—the place turned out to be a dump. The promise of paradise meets reality, an apt metaphor for Brazil at the time.

Altamira still underwhelms.

The city was assembled in a hurry. Freshly painted two-story concrete buildings alternate with weathered wood-slat shacks that slump as if too tired to stand up straight. The sidewalks are uneven, cracked, and overrun with weeds. As with most Amazon towns, music blares from storefronts, from cars with loudspeakers atop the roofs, from speakers on street poles, and from the boom boxes of those idling on their haunches watching traffic go by. Posters from past political campaigns, faded from the sun and rain, cover most of the public wall space. Shops selling car parts and electrical appliances, as well as a plethora of gleaming pharmacies, dominate commerce. Red dust-plumes swirl behind overcrowded buses, staining the buildings rusty-red. Most side streets are unpaved, and even the paved roads are rutted with potholes. Occasionally, a mule-drawn cart slows traffic, a reminder of how quickly this world has changed.

Altamira also holds symbolic significance in Brazil's indigenous environmental movement. In early 1989, ecologists, Indians, and politicians gathered there for a summit to discuss how *they* intended to manage the Amazon for themselves. Held shortly after Mendes's murder, the Altamira gathering attracted more notoriety than anyone had expected. That these various interest groups believed their collective decisions mattered was a revolutionary notion at the time. Mac Margolis wrote in *The Last New World,* "The summit also served notice to the policy makers. Gone were the days when soldiers ran things. Brazil's new democracy—and its financiers—would have to start listening to ordinary mortals now. Even ecologists and Indians."

The wide Avenida 7 de Setembro, named for Independence Day, is the town's main artery. As it leaves the thriving commercial center and works its way toward the quieter edges, it narrows, and smooth concrete becomes rutted stone and then choking dust or viscous mud, depending on the season.

Raimunda Socorro dos Santos Pereira lives at the end of this road, where wooden plank sidewalks perch over a garbage-strewn riverbed. The planks never meet, and they tend to sag. The wood groaned as we walked toward her home, vigilant for the visibly cracked pieces of wood. The shacks are built on tree trunk beams. Even the wallboards don't join neatly, so the rains can pour in. The high-water mark is about three feet up from the base of the shacks. There are months when Socorro and her neighbors live with flooded floors. Webs of wires run throughout the area. Some houses have meters, and the others "borrow" the electricity from them. There are no glass windows on the houses, although swinging board shutters cover some openings. There is no privacy, either inside the small warrens or elsewhere. Plastic bottles, rags, chicken bones, and crumpled wet paper accumulate under the houses. Green scum grows on the refuse, and the scent of steaming feces wafts in the air. Children play here while chickens peck at the ground and dogs scavenge. We saw two kids daring each other to dive into the refuse water under the houses. Mangy flea-infested dogs, too hungry to be dangerous and too listless to bother, are ubiquitous.

Socorro is twenty-eight years old. Never married, she has two sons. She arrived in Altamira from the Anfrisio River area about one hundred miles south of Altamira on August 15, 1994, after an eleven-day boat ride with her parents and sister. Socorro is thin with wavy black hair pulled back tight, and the day we met her she was wearing cutoff jeans and a tank top.

Wiry like a teenager, she shifted her weight from foot to foot nervously as she talked, unused to interviews about her life. "Life is better in Altamira," she said, mostly because the rural life had become so barren—with no schools and no health care. "Look, my mother not only could not read, she didn't even know what reading was," she said. "Now I can read a little. Maybe my children will be able to read. If God wishes."

Jacilene, Socorro's sixteen-year-old cousin, stood with her. She wore a T-shirt that advised "Try to Be Happy" underneath the picture of a dog. It wasn't working. Jacilene was sullen, clinging to Socorro. She had grown up on the Iriri River, also about one hundred miles south of Altamira, and had come here when she was ten years old. She's married with two sons, but she never attended school and can't read or write.

The cousins are from an area known as *Terra do Meio*—Land in the Middle. We heard several different boundary descriptions for this vast area, but it is roughly south of Altamira and framed by the Xingu River in the east, the Tapajos River in the west, the TransAmazon Highway in the north, and the large cattle ranches to the south. The Land in the Middle has been the site of continuous land conflicts between logging interests and settlers. The lushness of the remaining forest, the isolation of the area, and the long-term prospects due to its proximity to the feeder roads of the soon-to-be-paved BR-163 have made this area favored by the *grileiros* (land grabbers) and their hired *pistoleiros*.

C. R. Almeida, a businessman from the south of Brazil, the largest landowner in the country, owns (or at least controls) much of this land. Several activists told us that the *grileiros* operate aggressively because they believe that Almeida will be a willing buyer for any land they confiscate. Stories to this effect often appear in a quirky Belém journal written by Lucio Flavio Pinto, a courageous muckraking journalist who has nettled many vested interests in the Amazon over the past thirty years. Almeida's lawyers have repeatedly sued Pinto for defamation, primarily to force him to pay lawyers and to keep him in court and away from his beat.

Socorro took us to her house at the end of Avenida 7 de Setembro, where she lives with her sixty-nine-year-old mother, Dona Maria Raimunda dos Santos. Dona Raimunda's house has three rooms, each separated by a stained curtain. She sat on a worn couch framed by her grandson and daughter, like a weary boxer between rounds, splay-legged, wearing a faux-white-satin skirt and a white mesh top.

She brought her two daughters to Altamira because "the Anfrisio wasn't good anymore." Her husband had been a rubber tapper during World War II and its aftermath, a time of revival for the industry, when the landowners organized the workers into "rubber colonies" and provided amenities such as schools and health centers. When the rubber market collapsed again, these benefits disappeared. The tappers couldn't hunt for jaguar and the local jungle cats, whose skins were valuable for export, because a new law prohibited this activity.

Dona Maria Raimunda lost five of her eight children before they were toddlers, a reminder that heartbreak is also a debilitating disease. "One was stillborn, one died during birth, one had malaria and died, one died at two from a spider bite. Another died when he was nineteen months; he got a fever and was dead in three days. There was something behind his ear. It was bright purple and warm to the touch.

He was stung. There was no doctor." She left a son and daughter-in-law behind in the Anfrisio area, and two of their children died before they were three.

Nearby, eight children and five adults live in the house of Manoel Nazare da Silva's father. The children were grimy, in various states of dress. None wore shirts; some wore shorts made from worn adult T-shirts. One small boy had no belly button, just a mound of scarring like a beaten golf ball embedded in his belly. All the kids' stomachs were distended from malnutrition. They coughed and sneezed, and mucus covered their dirty lips. As flies swarmed, two women sat breast-feeding babies, each about a year old.

The entire house was one dank room, the only furniture a pair of wooden benches and a white plastic basket on rollers. The air, an almost solid, smelly, suffocating substance, did not circulate. Hammocks hung in the dark corners, and some shredded magazine pages and graffiti in white chalk decorated the walls. The rain clanged on the tin roof, and the house swayed on its wooden stilts.

Silva had come here to visit his sick father and hadn't been able to return to his home on the Anfrisio. He couldn't afford the boat ride back. He and his wife, Maria Valadares da Silva, have five children ranging from eight months to ten years. None of them has been to school. "There is no school on the river," he said. Every family there has about ten children, he told us. They live off trading Brazil nuts, rubber, and fish. They have no money. He said he doesn't like the city because "the situation [in Altamira] is not life." It is too crowded and busy, he said, and he can't grow his own food. Yet he feared that he would end up staying.

We walked from house to house, attracting stares because of our obvious foreignness. As desperately poor as the area was, we never for a moment felt threatened or unwelcome—just the opposite. Kids came out to hold our hands. People invited us to sit and have a *cafezinho* (small coffee) with them. Cordiality is a national trait, as Joseph Page points out in *The Brazilians*, no matter the circumstances.

—

The Live, Produce, Preserve Foundation (FVPP) is an organization in Altamira that works with the rural poor who flee to the city. Run by Luzia Pinheiro, thirty-five, and Antonia Melo, fifty-seven, the organization sees its mission as checking immigration into urban areas by providing services to people living in the forest. While they don't consider themselves environmentalists, Melo noted that "if the people in the forest are not respected, then they and the forest will be destroyed." She added that her organization provides support services to the new arrivals in Altamira, although by then it's often too late to help. "The rural people can't adapt," she said. "It's a different way of life, a different way of finding food, different sicknesses, different abilities necessary to survive. These people are much better off in the area they know, if only we can improve their lives there. These people need basic services we don't even think about. Many people don't even know what money is."

One of Melo's goals is to provide the river dwellers with an identity card, like a social security card, which is essential for everyday life in any town or city. "When these people have documents, they believe they are people with rights," she said. "It changes their perception of themselves. That is important."

According to her, only one woman who lives along the Anfrisio River has had formal education. "If they can't read the Bible, then they can't have religion. They can't read newspapers, so they don't know what is happening in the world." Health services is another of her priorities. "There is no hospital there. When people need medical attention, they often die on the route to get it."

The market economy of the river adds to the cycle of impoverishment, said Melo. It depends on *regatoes,* slow-moving ark-like ships that buy fish, Brazil nuts, and contraband skins from the river people and sell them dry goods in return. The *regatoes* buy cheap and sell exorbitantly high. Melo explained that "these boat shops will sell a pair of underwear that costs two reais in Altamira for fifteen on the river, because there is no other market. The people always end up owing something to the *regatão* because the exchange of fish and Brazil nuts for goods is not equal."

FVPP's goal is to somehow organize the diaspora of the river community, which is united by a network of blood relatives and by decades of intermittent communications. However, Pinheiro and Melo are discouraged by their inability to attract politicians' interest for their cause. Their constituency is too scattered and too disenfranchised to provide a bloc of votes or campaign contributions. Politicians can find more money, more votes, and many worthy causes in a square mile of Altamira than in 625 square miles along the Anfrisio River. Both women are resigned to there being a cycle of violence and poverty.

Without protection from law enforcement and without a basic understanding of their legal rights, the river people, because of their isolation, are prey for *grileiros*. Pinheiro said, "The loggers go in front, and the *grileiros* follow right behind. People actually show the *grileiros* around their land without knowing what would happen. The *grileiros* expel families from their homes, and some people leave to escape their threats. These people now live on the periphery of Altamira. Have you seen the misery that surrounds Altamira?"

Tarcisio Feitosa works with the FVPP as an activist organizing the river population to assert their ownership and squatters' rights to the land they occupy. Like the priests of a generation ago, Feitosa allegedly is a marked man. People told us it was just a matter of time before he would be assassinated.

Sitting in his house along one of the side streets of Altamira—really just a door in a wall leading to several unfinished rooms surrounding a well-appointed kitchen—we watched a slide show of maps of the *Terra do Meio* that highlighted the dangers he saw. He displayed an instance of deforestation occurring from August 2003 to June 2004. "It's an area of sixty-two square kilometers [38.5 square miles]. You can see the dot in August, and it's all gone in June. The government knew this and didn't act, or it acted too late." He looked up. "This is not just about trees, you know. Deforestation brings an accumulated violence against the population. Trees die, people die, too."

The salvation for the residents of the *Terra do Meio* may be that it has become so violent the government will have to intervene to avoid the embarrassing publicity that exposes its lawlessness. The example of civilizing forces eventually dampening the violence in the south of Pará doesn't apply here. "This is different," Feitosa said, "because all of the activity is taking place in a more concentrated region. In the south of Pará, even though there was violence in certain areas, it was a much bigger region so you didn't have these constant confrontations. In the *Terra do Meio* there is pressure all the time."

Feitosa, thirty-three, travels often in the region and has a network of informants, including a boat driver who arrived during our visit and delivered news about his trip. Feitosa also is a fixture at regional conferences dealing with the problem of violence and deforestation in the *Terra do Meio*, although he said he feels like the unwelcome messenger of despair. "The *Terra do Meio* won't last another five years at the rate things are going. The situation in the region is critical. The people are suffering. Soon they will have to escape into slums in Altamira or be destroyed along with the forest."

We invited Feitosa to dinner at a pizza restaurant along the riverbank. He asked if his wife and young son could come along. Then he asked if the boat driver could join us. Then he asked if the boat driver's family could join us. When we arrived at the restaurant, he asked if his priest could join us, as well as two women who just appeared at the table. We cynically suspected he was abusing our hospitality, but that wasn't the reason. "You see those men at the table over there," he said, and pointed to four men who looked like the cattle ranchers we had interviewed. We had noticed they'd turned their chairs toward us and stared at us when we walked in. "They have a contract on my life," Feitosa said. "They won't do anything here with our families, but they want to kill me. I know they do."

The priest, Father Robson, nodded. "They probably want to kill me, too," he said. "Tarcisio is a problem for them. So is the Church. We are organizing the people around the ecology and against those who are destroying it. They want to destroy us."

It was an eerie dinner. One of the ranchers would get up and walk behind our table, chatting on his cell phone, and Feitosa would spring from his chair, stare him down, and dial a number. The boat driver huddled his wife and kids together, all of them eating off the same plate. At one point a rancher walked across the street and disappeared into a building. Father Robson switched sides so his back wouldn't face that building. All the while, the conversation continued. Father Robson reflected on the explosive growth of evangelicalism. The Catholic Church couldn't compete for new worshippers anymore, he lamented. "But I am not interested in finding new people for the Church. I am interested in organizing the people we have. The evangelical church is for the individual. The Catholic Church is for the community. And if there are people who watch television and want to be evangelicals, there is nothing I can do about that. This is not a competition. It is about making lives better."

At the end of the night, Feitosa asked us if it would be all right for them to take the first taxi for himself, the boat driver, and their families. We stayed behind and saw them safely off.

A man named Pula Pula lives among another collection of shanties on the edge of Altamira, called Porto 6. His name means "Jump Jump," because he suffers from uncontrollable convulsions that shake his body every couple of minutes. It's like an electric charge running through him. He calls it *congestão*. Some have told him that *congestão* is caused by drinking hot coffee and then ice water without waiting long enough. His sister told him that women get *congestão* from working over a hot pot and then going out in a cool breeze.

We had come to hire him to ferry us by boat down the Xingu River. Squatting on the floor of the dark one-room house, sweating and swatting flies off his skin, Pula Pula assured us of his skills as a river pilot.

But how could he pilot a boat in the angry currents of the river if he couldn't keep his head from jerking spasmodically from side to side? If he couldn't control the movements of his arms, how could he steer? The Xingu River doesn't follow a straight line; it has multiple channels and hungry currents, all requiring strong navigational skills and instant judgment. No services exist to repair broken motors en route or to rescue capsized boats. A pilot alone determines a journey's success. He shrugged off his handicaps. "I know the river," he said. "I know its people."

The next day, wearing a hat that said PILOTO, Pula Pula squinted as his skiff slid along the turbid river. As Altamira and its detritus faded from sight, Pula Pula's spasms softened into twitches, as if the river were a kind of medicine. "My father was a Mundurucu Indian, and he spoke Guarani. My mother was from Maranhão. She came there with her father to look for rubber. My father died from malaria when I was four. I have three sisters and six brothers. I am the youngest."

One afternoon, Pula Pula told us, his father went hunting, and Indians who were not Mundurucu came to his settlement. They killed almost everyone and "stole" his mother. They ransacked the area, and seized the food and animals. When his father returned and learned what had happened, he gathered his friends and went after the Indians. "They rescued my mother," Pula Pula said. "It was an insecure, violent world then," he said. "But, more now. It's become more dangerous because there are gunmen, bandits, and *grileiros,* thirsty with greed. Because of FUNAI [the federal agency responsible for the Indians' welfare], the Indians are no longer savage. It's the white men who are now."

Pula Pula guided the skiff with confidence, veering around obstacles, mostly telephone-pole-like logs, and many others that only he could see (or sense). He would abandon one bank for the other for no apparent reason, or slow down to a crawl to cross an ordinary stretch of water. He'd explain why he chose one channel over another, but they all looked the same, and there were no discernible markers. He was almost like a bloodhound, guided by an undetectable scent.

The hours go by on the river, hypnotically and monotonously. The fishermen and open boat riders bake in the sun, no matter how refreshing the breeze. The natural beauty of the shoreline loses its luster after miles and miles. We spent many hours along the main arteries and tributaries of the Amazon, and all that changed were the currents and the colors and depths of the waters. The people and the scenery—thick forests occasionally gouged by thatched huts with smoking tree trunks behind them—vary, but not much.

Riding the river with Pula Pula was like walking a ward with a Chicago alderman. He stopped at every fishing boat to gossip, and slowed down before every hut to see if there was someone to greet. He circled the dugout canoe of Manoel Eladio Viana, an old man with dark wrinkled skin, callused and knobbed hands, and fingers stiff and warped from a lifetime of paddling and fishing. Viana wore a red flat-brimmed baseball cap, a horizontally white and black striped shirt that hung loosely, and bloodred sweatpants. His canoe, a hollowed-out tree trunk, held several inches of water in the bottom, and small silvery fish swam among his toes. After thirty-five years of soaking in bilge water, his toenails had fallen off. Gnarled lumps of skin had taken their place. Viana fishes every day by himself, and he sells his catch to businessmen who come by in boats with freezers. He sells *pacu* and *tucunare* for about one real. In Altamira they sell for four times as much.

We stopped to visit Seu Francisco Mendes, who, like Pula Pula and millions of other Brazilians, has a descriptive nickname. People call him Cu Queimado (or "Burnt Ass"), because they say most of the time he is drunk. Years ago, Mendes "messed with" a young girl named Eliana Kuruaya of the Kuraya tribe, and was arrested. He then waited until she was fifteen and asked her to marry him. She's twenty-six now, and he's fifty-three. They have six children, the youngest six months old. Mosquito bites cover his children's skin. The bites look like a rash. Invisible *piumes* attack persistently, leaving behind bites that bleed and itch for days, tormenting everyone living along the river.

Mendes was wearing no pants, just an oversize striped shirt down to his thighs. His hair and beard were white, although his mustache and eyebrows were still a nutty brown. A line of spittle gummed his beard from the center of his lower lip to his chin, but he didn't notice. He was emaciated, yet full of energy, constantly chattering at his children. He loves his wife, he said, and kissed her constantly—cheeks, shoulders, neck, arms—pecking like a hungry bird.

Dona Eliana's face was round, her skin was reddish brown, and her black hair was pulled into a ponytail. She spoke deliberately. Her husband, she told us, "has a good heart and works hard. I feel bad for him. I need to help him. He's more than twice my age." He plants cacao, manioc, watermelon, papaya, and banana, working even after a bout with a bottle of *pinga*. He fishes, too, selling his catch to the *regatoes*. He takes care of his seventy-three-year-old father. His six children are not starving. No one leaves his land without his insisting that they take food with them, papaya or manioc or fish.

Three more hours down the river, we met Seu Alfredo Carvalho Alves. His six children and their families live in various huts around a small clearing on Piranhaquara Island. There are fourteen adults and eleven children. Only Elaine, who is fourteen years old and who studied through third grade, knows how to read and write. The family arrived on the island twenty-two years earlier to become rubber tappers, but the rubber ran out. "We fish the whole day, sometimes just to get four fish," said Seu Iran Geraldo de Jesus, forty, one of the sons. "People don't earn a good living here. Sometimes we sell farinha. But I can survive here, and that's good. It's better than in the city. In the city there is rent, and there would be no money left over for anything else."

He called his house a "roof" and said it had to be changed every five years. They were changing it when we visited, so the clearing was covered with large drying palm fronds, in various hues from green to a bleached brown.

The family gathered in the largest structure for dinner, and as darkness descended, they became slow-moving silhouettes. Dona Maria Jose, one of the daughters-in-law, is very thin, and her stomach bulges. She's been in pain for over a year, but she said she doesn't have money to go to the "street" and be treated. When they have malaria they go to the health post in the indigenous village two or three hours away by boat, but the doctors there didn't know what to do about her stomach. Dona Linda, a daughter, said sometimes they try to sell chickens to passersby. They need money for sugar and detergent and oil. They don't know how to make soap or how to extract oil, and they covet instant coffee and processed sugar.

The children eat from plates on the dirt floor, grabbing at small clumps of rice and bits of fish as their mothers hover above them in darkness. "They are hungry," Maria Jose whispered. "Today the men didn't fish, so we don't have anything to eat." At night the children cry a lot; some whimper like wounded animals. "They are hungry," Maria Jose whispered. They are surrounded by trees bearing abundant fruit and live alongside a river that is an aquarium; yet their children scavenge in the dirt. How can this be?

About an hour away and for no apparent reason, the family of Fernando Dias Ferreira de Carvalho thrives, relatively speaking. On the Carvalho land, two large well-tended buildings sit on a long strip of cleared riverbank. On the day of our visit, laundry drying on the lines flanked the larger hut. The smaller hut is a comfortable house. Corn was drying on roof beams under the thatched roof, and the openness let in a cool breeze that made it difficult for the *piumes* to attack. There were columns of clean silver pots and pans. Ducks wandered around mischievously, not in the predatory fashion of the chickens we had previously seen. A baby sat on the floor grabbing corn mush while his mother, seated on a tree stump, shooed away the playful ducks.

Fernando Dias Ferreira de Carvalho, forty, and Maria Nazare Borges da Silva, have seven children, all still living. His seventeen-year-old daughter is married, and she has a one-year-old son.

He has lived along the river for twenty-seven years. Sometimes *grileiros* come and threaten people to get off their land, but he hasn't had that problem. He said he grows corn and manioc, and hopes to grow cacao soon. "Sometimes I hunt, and we bring back a large pig, which we can eat for one week."

"The situation here is sad," he said, unaware of the even sadder plight of his neighbors down the river. "Nothing we buy from the *regatão* is cheap. They give me only 1.70 reais for a kilo of good fish, and I have to pay five reais for a kilo of sugar. I fish every day. I have to." Carvalho lamented the stranglehold the *regatões* have over river commerce. He was dismayed, of course, that the fish he sold were marked up three or four times when they were resold in Altamira. He thought it a crime that manufactured products from Altamira were marked up seven or eight times for sale along the river, with the higher prices reserved for those living farthest away. Those who can afford it the least are those exploited the most.

The homestead of Carvalho was so tidy that we barely noticed the floor was dirt and that there was no electricity. His child's laughter provided a signature sound to the home. His nearest neighbor, meanwhile, wallowed in filth. It wasn't just that his neighbors were newcomers, or that they lacked help. The river people have created a network, even a far-flung neighborhood, and they work each other's land when necessary. Yet with two families living in the same environment, one succeeds and one never will.

This disparity hung with us throughout our visits. We were seeing contrasts that Jared Diamond in *Guns, Germs, and Steel* had characterized as "the innate differences in people themselves." This was a much more difficult inquiry, we thought, than his study of "different peoples because of differences among peoples' environments, not because of biological differences among people themselves." Why was Pula Pula so inquisitive and so energetic, while his contemporary, Manoel Eladio Viana, fished silently every day until his toenails fell off? Why had Socorro gone back to school at age twenty-eight, while her cousin showed no interest? Why were we seeing such different conditions in the river people so proximate to each other? We wondered if there are answers to these questions, and, if so, how does a country encourage the positive qualities and confront the negative?

—

Along the way, Pula Pula introduced us to a *regatão* pilot, Antonio Neves Alves, forty-six, who has run the river for almost twenty-five years. He spends up to two weeks a month away from his wife and six children in Altamira, bartering for goods to bring back to the city. While he's sympathetic to the plight of his customers, he, too, must make a living. "Everything here costs a lot more. What costs twenty-five reais in São Paulo costs three hundred reais in Altamira. What costs three hundred in Belém costs six hundred here." Alves pointed out that without his commerce, the river people would have no access to basics such as soap, canned foods, or clothing.

The *regatão* system—now denounced regularly as exploitative, since people know where prices stand elsewhere—was once seen as a liberating economy in the Amazon. Prior to the *regatão* system, most river dwellers were subjected to a monopoly controlled by large landowners, who made goods available to their workers on credit. This credit had to be worked off, and the effect was that river dwellers were caught in a kind of enslavement. The arrival of the *regatão* driver broke this monopoly and exposed the river people to market choices (as long as there was competition among the *regatões*). David McGrath of IPAM wrote an article on the *regatões'* commerce, noting, "Some writers have argued that they play a destructive role in Amazonian commerce, cheating unsuspecting *caboclos* [native Amazonians] and stealing the customers of legitimate businesses, while others maintain that they have played a progressive role, introducing an alternative to traditional commercial relationships based on debt peonage." McGrath describes the *regatão* system as embodying the "Janus-face of merchant capitalism."

Here was another example of the far-flung repercussions of the forced exodus of the river population to the cities. When the rural population thrived, as during the rubber boom revival in the middle of the last century, there was healthy *regatão* traffic and ample market choices. Now, with the rural population dwindling because so many are moving to urban areas, merchants ply their trades where the customers are. Those left behind in the forest don't present a market that will attract competitors, and a single *regatão* pilot will control a widespread area. There is nothing inherently exploitative in the *regatão* economy; the prices function in relation to supply and demand.

In the clearing known as "Go Whoever Wants To," we came upon a satellite dish, a freezer, and a latrine, the home of Anastacio da Silva Avelino and his wife, Iraildes. Here global consumerism had found a place along the Xingu River—there was a television, a sound system, a CD collection, and even fake plastic flowers. Large fans circulating in the two main buildings kept the *piumes* at bay. There were doors that actually opened and closed, glass windows, and lights. The bananas were shiny yellow, not sadly spotted and rotting, and there were shoe boxes filled with fresh eggs.

Avelino's skin resembled burnt leather, contrasting starkly with his powder white hair. He had a stubble of a white beard, a mustache, and a broad, weathered nose. An egg-shaped bump had hatched over his left eye. He was barrel-chested and dressed in a clean aquamarine shirt that hung loose and open over his swimming shorts and ragged flip-flops. His left middle finger had once broken, and then had been fixed at a strange angle.

The Avelinos have twenty-one children, seventeen of whom are still alive. Two sons, ages twenty-eight and twenty-five, live with them. Avelino's five-year-old grandson, Ze Preto (they call him Joe Black because of the color of his skin), also lives there; Ze Preto's mother ran off one day and never returned.

Avelino, born in 1932, had been a rubber soldier during World War II and its aftermath, and he collects a union pension. He was a boat pilot and a farmer. With his savings, he bought land downriver near the town of São Feliz do Xingu. He said that about six years ago a FUNAI representative, Sr. Dimar, told him that he and his family "occupied indigenous lands and would have to move. I said no. I said my family lived on this land for forty years, and I was certain it was not Indian land. So Sr. Dimar told me he would give me an indemnity of eight hundred thousand cruzeiros for the land and the trees. But he did not pay me. In 2000, I made an official document that said my family occupied the land before it was protected land for Indians."

Avelino bragged of vanquishing the bureaucracy. But then the *grileiros* arrived. "João Cleber, he's a *grileiro* and a *madeireiro* [a logger], and his *pistoleiro* José Daniel came to bother me. I was having problems with João Cleber for a while because he would cut down mahogany trees on our property and not pay us. I made a statement to the police in São Feliz, but they did nothing, so my family went without, just sucking their thumbs."

Avelino would not move. "José Daniel told me I would have to leave. On July 30, 2003, he came to my house with many other *pistoleiros*. They shot up the house, and they shot the furniture, and they killed the animals. My boys were hiding from them." Avelino and his wife were in Altamira. "We received the message on August 8. My sons Lindomar and Devanildo told me. They said that Manoel, my nephew, was crazy with fear and ran for the boat when the *pistoleiros* still were there, and they shot him in the leg."

As Avelino told the story, Dona Iraildes shuffled behind him with her hand over her mouth. She yelled over his shoulder, "Do you see? Do you see what they did?" He paused and let the outburst pass.

"Why would these men do this?" Avelino asked. "I've never taken anything from anyone."

They left Altamira and traveled three days by boat to São Feliz do Xingu and made a report to the police. The police asked for three thousand reais. "I told the police they were paid a salary by the government. But I know the *grileiros* were paying more," he said angrily. "I refused to pay a bribe, and I knew I would lose. I would lose if I paid them, because the *grileiros* can always pay more."

Dona Iraildes cried, "The police are bought!"

"João Cleber demanded that we be killed," Avelino said. "They killed a neighbor of ours. Levi died. An old man. He was working his clearing, and they shot him in the back. Why? For money. For land. I may be illiterate, but I know this in my gut: they did this just for money. What is this world we're in coming to?"

His wife covered her mouth. "They could have killed my boy. They still could."

When they returned to their home at Bom Jardim, they found bullet holes in the hammocks and the furniture. The possessions that hadn't been destroyed, Avelino claimed, had been stolen by the Indians.

"The Indians," Dona Iraildes declaimed. "We give them manioc when they are hungry because they are so thin. They came to us so thin! And I gave them food, and they stole it all."

Avelino bought the land at Go Whoever Wants To locality when the FUNAI deal was struck. Now he determined to move his entire family there and start over. "Someone who was sent by João Cleber gave me money for the land. I never went back."

"But maybe they will come here," Dona Iraildes said. "The way they did there. We don't steal. We fight to earn what we have. But the big and powerful steal from the poor."

Francisco da Costa, who lives on the other side of the river from Avelino, has his own set of problems. He has no electricity and none of the other comforts of the modern age. He has a small hand-dug canoe and uses it to catch fish, which he trades with the *regatão* for olive oil, manioc, flour, and soap. His grandparents settled in the area after fleeing the perpetual drought in the state of Ceará. Da Costa is forty-four years old, married to Francisca Teixeira Moraes, thirty-six, and they have six children, the oldest being twenty-two with two girls of her own.

"Life here is hard," he said to us. "Some days there's food, some days there isn't." He is a short man, his rectangular face framed by unkempt wiry salt-and-pepper hair. His eyes are dark and liquid, sorrowful. He wore bathing trunks, no shirt. His family, swarmed by *piumes*, listened as he spoke. His three dogs listened too, filthy, sickly, covered with grape-size ticks, and surrounded by a haze of flies.

Da Costa complained that the fish population had dwindled and that it was difficult to get a worthwhile catch even after a day of trying. "I live a kind of humiliation," he said. "I don't know how to do anything else. I trade with people who have other things. I don't have anything. I want to plant, but I don't have the resources to do that."

But then he added, "There are people here on the riverbank who really don't have anything, not a thing."

Two years earlier, his house burned down. "We lost anything we had. Even our official documentation. Things like this happen." He shook his head. "We are here only by the grace of God. The powerful one above sustains us."

We headed to São Feliz, the largest town on the river south of Altamira. It was time for Pula Pula to go home. We thought of the contrast between the Pula Pula we had first met, the sweating, twitching figure in the stifling house in Altamira, and the Pula Pula we were leaving, a man comfortable in his environment and welcomed everywhere along the river. Pula Pula needed the river, because it washed away his illness. It gave him a sense of purpose, which was the common trait we perceived among the successful settlers we met. None had any sense of self-celebration, only a seriousness applied to the task at hand, as well as an absence of self-pity, the other defining characteristic of those who'd succeeded. In a setting where the goal was to clear land, build a house, feed a family, and keep it healthy, to do so was a major accomplishment, something not evident until you noticed how many people lived without.

Celebrities and their glitz, the staple of Brazil's mass media, provide an escape from this reality but little in the way of role models. No soap opera writer would deign to focus on the mundane heroism of Pula Pula or the river citizens we met, as there's no cachet in the basic fight for survival. Recognizing the women who run the FVPP and these women's counterparts, and their quest for universal education and access to health care, might provide inspirational role models, but recognizing them would also provide reminders of an unwelcome underbelly of society. Pula Pula and his friends along the river certainly have the character to succeed. Almost 100 years ago, Theodore Roosevelt observed this, too, when he wrote, "[One] could not but wonder of those who do not realize the energy and the power that are so often possessed by, and that may be so readily developed in, the men of the tropics." If only they were given the very basic tools of development— an education and the chance to live a healthy life.

Pula Pula would run his skiff all out under the navy gray overcast sky until he arrived back at Dona Teresinha's, because he knew that if he stopped there, it would not be an imposition and she could tell him about the stories in the newspapers. He would find an unused hammock and some space in the big hut and nap. He would tell stories with Dona Teresinha while she cooked a wild boar and then insisted on his taking strips of it, wrapped in leaves, to fortify him during the two days it would take him to get home to Altamira. They would entertain themselves by carrying on a conversation with Lourinho, the parrot she's had for seventeen years, and the two of them would lead the bird around the hut by dropping a trail of rice kernels. They would laugh about how smart the bird was and praise each other's good health, and then focus on the people struggling for life along the river and try to guess who would die, or be kicked off their land, or both. This conversation would make Dona Teresinha tremble, almost like Pula Pula. Soon she would say, "Enough" and they would talk about President Lula and the soap operas, and the time would pass. Pula Pula would take his leave and ask Dona Teresinha what he could bring to her on his next visit, and she would say, "Nothing." But he would think of something. They are friends, and they look out for each other.

CHAPTER 16

LAND, VIOLENCE, AND HOPE

—

Issues of landownership and the accompanying violence occupy a middle strata of turmoil, between the impressive economic enterprises of cattle and soy and the hopelessness of the displaced poor. Nearly every large tract of cleared land in the state of Pará, if not also in the neighboring states, has been the site of some violent struggle over occupation or ownership. Economic victories have left victims by the wayside. As in Altamira, the victims, who arrive expecting a better life, end up impoverished in the cities. Or they move deeper into the forest, expanding the frontier, only to become targets again in a few years. The victors consolidate the spoils, often legitimizing their acquisitions by using corruption, lawyers, or both, and then they provide legal title to real businesses. This situation may not even change once the frontier closes. Unless the government can impose adequate physical security and civil law over this massive region, which is a doubtful prospect, given the competition for the government's attention elsewhere, the underclass has slim prospects for upward social mobility.

The United States, whose frontier was as expansive as the Amazon's, had multiple temporal and substantive advantages that allowed for the success of that frontier's settlement. Few of these advantages exist in Brazil. The American frontier opened in the nineteenth century at a time when long-haul transportation was generally unavailable or slow; this impeded spontaneous diasporas. The flow of pioneers to the American frontier was manageable, partly because there were acceptable stopping points along the way, such as the Mississippi River towns. The settled eastern part of the United States wasn't overcrowded and spewing out people fleeing from drought; the spirit of our pioneers was more optimistic, less desperate. Traditional conditions for emigration from settled places—famine, unemployment, landlessness—hadn't risen to anything resembling crisis levels in the United States. The decades between the discoveries of Lewis and Clark and the frontier's opening allowed the government to establish a law enforcement presence, as well as the semblance of civil society in the form of municipal governments and outposts for education and health services. Almost all of the land belonged to the federal government, so land titles transferred with definition and certainty.

The Homestead Act defined America's frontier, as it allowed for the transfer of ownership to those who agreed to make the land productive. At first, the landholdings were modest, consistent with the government's goal of planting a middle class in the middle of the country. The viability of the legislation lay in the government's ability to convey recordable land titles, because it could draw maps showing measurements and boundaries. Land inventory was certain. It could be cleanly and clearly owned.

Had settlement of the Amazon frontier proceeded similarly, nearly all the problems associated with it—deforestation, violence, urban poverty—might have been ameliorated. The settlers would have had vested interests in protecting their communities and environment. But Brazil's settlers arrived on the frontier practically overnight, outrunning the arrival of physical infrastructure, health and educational services, a civil law system, or any law enforcement structure. The Amazon frontier essentially opened for business when Brasília was completed, which in turn led to the construction of the Belém-Brasília

Highway, completed in 1960. (Manaus and Belém on the Amazon River were already established cities, but without any road links to the region, they never served as a launchpad for any significant migrations.) Most of the land-tenure violence over the last thirty years has taken place within a few hundred miles of that road. Ironically, Decree Law 1164, issued in 1971, gave the federal government ownership of sixty-two miles on either side of federal roads, and the government failed miserably in governing those tracts. In the late 1960s and early 1970s, the three-thousand-mile TransAmazon Highway was built. The mass media campaign promoting settlement around the Trans-Amazon Highway in the 1970s created uncontrollable demand, because the propaganda promised there would be ordered public colonization projects with adequate support services. Nothing like this appeared. In *Contested Frontiers in Amazonia,* Marianne Schmink and Charles Wood quote a gas station attendant along the highway who describes it as "a poor man's road. It links nothing but poverty in the Northeast to misery in Amazonia." Within ten years of the opening of the Amazonian frontier, there was access to the region through major routes running east to west and north to south. Demand for a second chance in life had built up in the settled parts of Brazil—in the parched northeast, where the promise of a lush Amazon was alluring to people condemned to perpetual drought, and in the agricultural south, where tenant farmers were seeking their own land.

As no preparations had been made for this movement, chaos ensued. The only other instance of large-scale Amazon forest clearing followed by human occupation had been the Madeira-Mamoré Railroad, which had unleashed malaria and a menu of tropical diseases. The same happened here. Health care wasn't available. A sense of bewilderment settled over the frontier areas, as there were few identifiable signs of civilization. Just roads, clearings, and rudimentary living quarters.

Most important, there was no reliable system of land tenure. Titles dated back to the reign of the Portuguese, and most titles described tracts of land so large that no certain measurements existed. This was a time before satellites and GPS devices were available for mapping; no one had expected land surveyors to trek out into primary rain forest to stake off plats. Moreover, these older titles often differed in what information they specified, which led to more confusion. Some titles conveyed rights to extraction of trees, while others conveyed the land itself. (Texas has a similar structure, with distinctions between rights to explore for oil and gas versus fee simple ownership.) Many of these documents would have required forensic experts to distinguish between the valid and the forged. In *Passage Through El Dorado,* Jonathan Kandell writes, "By the 1970's, an exasperated governor of [the state of] Acre complained that if all land claims were honored, his state would have to be five times its actual size."

Certainty in landownership is an asset every productive society must have. The economist Hernando de Soto observed in his native Peru that "no resident of an informal settlement will invest much in a home if there is no secure ownership of it, no street vendor is going to improve the environment if eviction is feared, and no minibus operator will respect public order on a route to which rights are not recognized." De Soto's concerns echo those of the economist Ronald Coase, who observed the interdependence between private ownership of land and the health of the environment. According to Coase, "many disputes over resources stem from the fact that no one owns them. Or— nearly as bad—everyone owns them, as in the case of public property. However, these disputes could be resolved if the unclaimed resources were divided up as private property. Now if someone wants to use your property, you could charge them a fee. Or if they abuse your property, you can sue them in court. Assigning property rights greatly enhances the ability to resolve disputes over the use and abuse of resources." Accordingly, assigning property rights also holds people accountable for the use of their property, enhancing a government's ability to control development.

Daniel Webster recognized the interplay between private landownership and a stable society in a political context almost one hundred years before these economic theories evolved. He argued for a wide diffusion of property ownership as the basis for government's credibility: dispersing property among the greatest numbers invests more people in the institutions of government, resulting in respect for them. Webster feared the effects of an accumulation of property in the hands of a few, and these property owners' resulting controlling interest in government.

Landownership theories address a sense of belonging, which has been as elusive on the Amazon frontier as it was where the settlers came from. The absence of this sense has largely defined the development of the Amazon. Who would build a house on land he did not clearly own? Who would plant? Who would plan for passing on a legacy? Who would feel safe knowing that others claimed ownership of that same land? And as this land was often surrounded by inhospitable jungle, by armed marauders who bought and sold the police, and by hungry entrepreneurs intent on increasing their landholdings, a stable civil society was impossible. With this uncertainty, communities could not grow. Public and private investment would be transitory, as no potential long-term returns could be observed. Communal security and the rule of law flow from the creation of a system that allows for the identification and protection of property rights. And, remarkably, as de Tocqueville observed in the developing United States, respect for and enforcement of the law in this civil society both emanate from below, making the system acceptable and effective. He wrote that a citizen complies with the law "because it is his own, and he regards it as a contract to which he himself is a party."

Where there is a lack of public authority and of shared investment in a community's well-being, might makes right. From these Amazon migrations, in which the status of the settlers was as uncertain as transients', arose an outsize slash and burn economy that has lasted for decades in many areas. It still persists at the edges of the frontier and into areas such as the *Terra do Meio*. If no one owns the land, no one respects it. Where deforestation and exploitative occupation are the only means of asserting practical ownership over land, no living thing is safe.

—

O Estado de S. Paulo, Brazil's leading newspaper, recognizing the state of violence due to land conflicts in the south of Pará, has called the municipality of São Feliz do Xingu the place "where the law values nothing and death costs 100 reais." On the edge of the settled frontier, São Feliz, with its sawmills and access to the country's road network, has long served as a gathering spot for mahogany loggers and itinerant miners who experienced a series of rich finds nearby in the 1970s and '80s. In addition, São Feliz is the nearest town to the villages of the Kayapo tribe, known for its willingness to defend itself. In 1980, the Kayapo killed twenty people, including women and children, who had settled on their land. An observation by Schmink and Wood in their study of the town in *Contested Frontiers in Amazonia* identifies the rest of the cast that makes the region so violence-prone: "For migrant families who traveled the PA-279 [the state road] in the hopes of finding a plot to cultivate, this meant that São Feliz was, literally and otherwise, the end of the road. After being evicted as many as three or four times from lands they had claimed, many arrived in São Feliz only to find that their prospects of becoming a landholder had hardly improved."

It's not an exaggeration that *everyone* we met in the town of thirteen thousand had been shot, kidnapped, or threatened, or at least had a close friend or family member who had experienced one of these violations. Schmink and Wood concluded after visits to São Feliz in 1978, 1981, and 1984, "It is no wonder then that the cruel realities of the frontier prompted migrants, Indians, miners and *caboclos* to engage in the increasingly organized forms of resistance.... As victims of violence and all manner of disappointments, their desperate struggle to survive in the face of ever more limited options was a potent force that redefined strategies people pursued to contend with the rapidly changing economic and political conditions in the Amazon."

Unlike Rendencão in Pará and Rio Branco in Acre, which have both progressed one hundred years in the twenty-five since our last visit, São Feliz has suffered stilted growth because of persistent violence. Discussions of killings are matter-of-fact, like the weather. People just shrug their shoulders, perhaps mentioning the number of bullets or the price paid to the *pistoleiros*. Perversely, the violence is also a barometer of the town's future: the *pistoleiros* only operate in markets of appreciating land values.

The archivist of much of this violence is a seventy-year-old American, Russell George Clement, whose worldwide wandering ended here in 1986 when he fell in love with a local woman. Clement owns the Xingu Lodge Restaurant, the only place in town with fresh tablecloths. Pictures of his adventures adorn the walls—fishing in India, tribal dancing in Africa, on the beach in New Guinea, and in the mountains in Nepal.

Most days he fishes, chats with neighbors, and watches what goes on from his restaurant. Originally from Ord, Nebraska, he still has a sturdy farmer's build and weathered hands. His daily uniform is a baseball cap, a sleeveless shirt, and a pair of jeans. He's the only American in town, which makes him as close to a tourist attraction as São Feliz has. The locals call him Clement, which bothers him. "It's rude," he told us. "Even my son's mother calls me Clement. It should be Russ or Russell, or Mr. Clement."

Clement himself was once kidnapped. "It was Sting's fault, really," he said. "It was 1994. Sting had come through here before that. He gave fourteen thousand dollars to one of the chiefs in one of the Indian villages. He didn't know what he was doing, and he's a bad man for it. All he wanted to do was increase his own publicity." Clement had evidently told this story before, as his sense of drama suggested. He stood up, paused, sat down, lit a cigarette. He was in no hurry. He was happy to speak English. His Portuguese was so mangled by his Midwest accent that few of his neighbors could understand him.

In 1994, Clement was visiting an Indian village with two American tourists who had come for adventure fishing. "Once we got there, the Indians wouldn't let us leave. They kept asking me if the tourists were from England. Kept saying 'England' over and over. Finally, I realized they thought that since Sting had given money to one village, that we would give money to them."

The Indians agreed to let Clement go back to town to get the money, but they held on to his visitors. "I came back to town, got my brother to wire me nine thousand reais, and they let the guys go. The Indians really didn't know what they were doing. It's all that rotten Sting's fault. Later, when I was gone one day, Chief Bry's wife sat in my doorway with her head bowed for hours. They felt ashamed. That was how she was saying she was sorry for them. We're all friends now. They come and visit in the restaurant. Pikanu is from that village, and he's a good man. He named his son Clement after me."

The relationship between the neighboring tribes from the Kayapo family and the São Feliz townspeople illustrates the cultural and economic tensions often seen in the Amazon. The Kayapo first had contact with Brazilian society in the late nineteenth century and, remarkably, have managed to hold on to an identifiable culture over a span of time during which scores of other indigenous tribes have become extinct. The Kayapo's belligerence has helped them, as they have resisted invasions of their land by loggers and miners, and they have shown an adroitness at managing the bureaucracy that the Brazilian government has established to govern Indian affairs. In 1978, FUNAI defined an area of 2,738,085 hectares for their reserve. The Kayapo responded: "Not enough."

Their concerns evidently included more than greed. They sensed the creeping influx of immigrants, the avariciousness of loggers eyeing their mahogany, and they espied the miners who had found gold on their land. Their 1980 massacre of twenty settlers stirred anti-Indian sentiment in São Feliz, but it also prompted the government to engage the Kayapo in the process of planning out some form of coexistence. Eventually, the size of the reserve was expanded to 3,262,960 hectares, which provided enough buffer area to deter invaders. In addition, the Indians negotiated a commission on the amount of gold the miners extracted.

Nonetheless, the juxtaposition of so few Indians (about two thousand in 1985, when they received the additional land) living so close to migrants so hungry for land, continued to breed resentment. Throughout the Amazon, griping about these set-asides is commonplace. The complaints are not necessarily racially based but are denunciations of an unproductive allocation of valuable resources. On the other hand, environmentalists delight in the Indians' good fortune, as there is very little deforestation on these lands.

That so much land was taken out of circulation in São Feliz has exacerbated conflicts. Clement confirmed the systemic violence. "All varieties, all the time," he said, although he has detected an abatement in recent years, which he attributed to the maturing of the town. "But still, land invasions are common, and people get killed. Once these *grileiros* receive the first payment for the land, they send in some *pistoleiros* to kill the guy, and they sell it again. Or sometimes to sell a large parcel of land, they send in the *pistoleiros* to clean out the area. Just three months ago, eight bodies were brought to the hospital. Once in the 1980s, a wounded man was killed in the hospital. Some *grileiros* had tried to do away with him, but he survived. When he was in the hospital, they sent a bunch of *pistoleiros* in. They just sauntered in, opened the door to his room, and shot him to death."

As a foreigner and as possibly the town's oldest citizen, Clement told us he felt that no one wanted to bother with him anymore. "They used to threaten me, because I wouldn't obey their code of silence. If I saw a murder, I talked about what happened and who had done it. No one talked, but I did. One time on the street, I heard from one of these criminals that the price for me was fifty thousand reais." Clement paused, stood up, walked away, turned around, and then patted himself on the chest. "I looked at him. I'm proud of this. I shouted, 'Only fifty thousand!' " He laughed and laughed.

José Gomes, nicknamed Zé do Largo, is one of those who was nearly murdered. In 2002, on his way back from a classroom, he found a man with a shotgun waiting for him outside his house. "He shot me in the chest, just below the left shoulder. You can see where the lead pieces settled—all seventy-two of them." There are scars on his neck, on his chest, close to his heart, on his arm and shoulder. The badly healed skin looks like peanut brittle. "The guy said I was messing too much in his buddies' areas, talking too much."

He still considers himself to be living on borrowed time, although he lives modestly with a wife and four children. "I'm a troublemaker, I know that," he said, although his work seems quite benign. He works on an environmental project, trying to repopulate turtles along fished-out streams. Zé do Largo said he assists denunciations that settlers make to the government about false land titles and corrupt police. That puts him in the most endangered class of residents of São Feliz. "I'd say one percent of the deaths here are from natural causes. The rest is murder," he explained. "There are four reasons people are killed here. The biggest is land. The conflict over land or trying to steal land is why people are killed. Politics is two. Wagging tongues is three, and caliber 48 is four." Caliber 48 are hired murders.

After he was shot, do Largo remained defiant in supporting his friends' causes but changed his focus from working with environmental NGOs to working with children and organizing noncontroversial events such as beach cleanups. He explained that he was not only safer in this job but probably more effective. "I started working with adults, but I wasn't succeeding. Men act like children—abusing and wasting. So I began working with children. They're more willing to change their ways and see the merits of new things."

In Maraba, the nearest large city to São Feliz, José Afonso at the Pastoral Land Commission (CPT) catalogues instances of violence and slavery in the area. His small office is lined with file cabinets that contain folders of information on every farm that has been implicated in slave labor, and folders of his reports on murders, death threats, and disappearances. According to his records, between 1986 and 2002, 386 rural workers were killed in Pará. In 2002, he recorded sixteen murders, five disappearances, 36 workers threatened with death, and 118 farms illegally "employing" 4,336 workers as slave laborers.

During our initial visits twenty-five years earlier, the Catholic Church catalogued these statistics and monitored these land wars. The CPT representative at the time was Father Ricardo Rezende, who, when we visited him, was monitoring almost one hundred disputes involving forty-five hundred families and eighteen thousand people. At the time, the government considered Rezende a dangerous man, not so much for his ideas—only slightly more radical than those of many other priests—but because he carried them out. Rezende did not just preach from a pulpit; in fact, he rarely preached at all. He organized.

What we remembered about Rezende was his unshakeable certainty in his cause, despite Pope John Paul II's admonishment to priests not to promote class warfare. Rezende was a child of the liberation theology movement, an outgrowth of the Second Vatican Council, which guided the Catholic Church in South America through the late 1960s and into the '70s, a philosophy that often pitted itself against the authoritarian governments on the continent. Liberation theology combined faith and politics. It advanced the view that Christ's becoming flesh meant more than a message of eternal salvation, that he did so to free mankind from the shackles of poverty and the humiliation of oppression. The message of the Latin American bishops at the Medellín conference in 1968 was that the Church listened "to the cry of the poor and [became] the interpreter of their anguish."

Brazil's clergy were among the leaders of this religious movement. One of its most visible proponents, an influential theologian, Leonardo Boff, wrote in 1981: "The Church is directed toward all, but begins from the poor, from their desires and struggles." Brazil's clergy, especially, embraced this view so tightly that it became the best-organized opposition party to the military government. Rezende echoed these beliefs when we met with him. "You cannot preach the gospel to a man with an empty stomach," he said. "He must have basic human dignity, and in this part of the world that means a piece of land where he and his family can grow food to live on."

The rural workers' unions and kin civic organizations throughout Pará today that address disenfranchisement issues can trace their lineage to Rezende's efforts. Rezende perceived disillusionment among the settlers and alienation from the nascent institutions on the frontier, which made the new arrivals vulnerable to organized and well-heeled interests. "This *is* class war," Rezende told us, referring to the assertion of the squatters' rights against the forged titles and hired guns of loggers and ranchers. "The Church is the most important force for change, and we must help people take the fight into their hands."

He explained, "Where the Church is most effective is in bringing the rural workers together, helping them to realize they have common problems and that together they are strong."

At the time, the pope sought to remove priests from politics, fearful not only of violence but of overpoliticizing religious messages. Rezende and his colleagues resisted. In 1982, two French priests in a parish near São Feliz were convicted, along with thirteen settlers, of "subversive" activities after the death of a government surveyor who'd been participating in planning their eviction from land on which they were squatting.

Rezende said he was tired of fighting guns with words, that unless the situation got worse, it would not get better. He said, "I will not tell someone to kill, but I will not tell him not to. These *posseiros* have suffered as much as they can. They have been moved off their land eight, ten times. But they will move no more. We will give them legal help. But Brazilian justice does not favor the common man. It would be possible for a lawyer to win, but it takes so long, four or five years. Also, there is tremendous corruption. There is sometimes only one alternative left. I cannot be the one to say, 'No, you cannot use it.' "

The world eventually changed for Father Rezende and the Church, but the tension in the state of Pará stayed at the same level, though it went deeper into the forest. Pope John Paul II meant what he had said and eventually dismantled the Church's political mission. His choices of Brazilian cardinals and bishops made sure that religion and politics would not intersect. Rezende now lives in Rio de Janeiro, having been transferred by a Church that was intent on removing itself from political causes.

Sister Dorothy Stang, a nun from the Sisters of Notre Dame de Namur, turned a deaf ear to the Vatican's directive to step back. Born in Dayton, Ohio, in 1931, she moved to the northeast of Brazil in 1966 as a missionary and settled in the Amazon town of Anapu in 1982. A press account of her life reported that she "lived among those who wanted her dead." On February 12, 2005, she was murdered.

We never met Sister Dorothy. We tried to travel to Anapu from Altamira, but the road had washed out. It would have taken at least six hours to travel the sixty-two miles, but the federal police told us, "You might not make it, and if you make it, then you might not make it back." She was murdered in a small town called Boa Esperansa, twenty-five miles from Anapu. The federal police in Altamira told us "it is not possible" to travel there during the rainy season.

The murder of Sister Dorothy has been compared to that of Chico Mendes—an international story of tremendous significance for the environment. It was fitting that the first federal police officer on the scene had been guarding Mendes's protégée, Marina Silva, at a conference in the Amazon River town of Porto do Moz when they received news of Sister Dorothy's murder. Silva, the environmental minister, announced the death and then boarded a helicopter and traveled to Anapu.

Anyone who has traveled the Amazon knows the regularity with which violent death appears. It's the unacceptable cost of the expansion and closing of a frontier. But Sister Dorothy's murder was different, and it was not because she was American. Here was a seventy-three-year-old woman who had lived without electricity, running water, and sewage. At a time when our own mothers were simply happy to have their health and the company of their friends, this woman literally walked hours, even days, from settlement to settlement in the sweaty mosquito-infested air, encouraging small farmers who had goals no loftier than feeding their families. Sister Dorothy was thin, with close-cropped gray hair, glasses, and a gap-toothed smile. She regularly wore a T-shirt with the slogan, "The Death of the Forest Is the End of Our Life." According to her brother, she had "set up a stabilization program that organized small farmers into cooperatives. The program included schools for the kids and workshops for the adults—child care and hygiene for women and crop rotation and sustainable farming for men." She had done this work far from cameras and reporters, without self-glory. Most significant for this deeply religious woman, she had championed this cause in defiance of the word of the pope, which was a statement of her devotion, not her rebellion.

On the morning of February 12, Sister Dorothy awoke in the settlement of Boa Esperansa, where she had traveled to meet with farmers she visited periodically. She had intended to leave the previous day but the meeting had run over, and she hadn't wanted to travel back to Anapu in the dark. She had come to Boa Esperansa to address conflicts between the farmers in the area and a rancher named Vitalmiro Moura, who claimed title to the land many of them were farming. According to Tarcisio Feitosa, Sister Dorothy knew there was a "contract on her life." He sensed that she knew that she would be murdered, that it was only a question of "when, where."

She stayed the night in a shack owned by one of the farmers. Ulisses Tavares, a federal police officer involved in the investigation, told us, "Wherever she wanted to stay, she could. People welcomed her wherever she went. She was very well loved. But she was also a rock in the shoe of the establishment."

At seven in the morning, Sister Dorothy and one of the farmers walked to where the farmers had gathered to continue the meeting. Officer Tavares recounted, "On the way, she met the two killers and they started asking her if she was afraid of dying. They asked if she walked around with a weapon. She stopped, held up her Bible, and said, 'This is my weapon.' "

According to the eyewitness, they peppered her with questions about what she was doing, whom she was seeing, but she ignored their taunting, intent on finding a passage in the Bible from Beatitudes: "Blessed are the peacemakers: for they shall be called the children of God."

"She knew then she was going to die," said Tavares.

Rayfran das Veves Sales, known as Fogoio, allegedly pulled out a gun. "Sister Dorothy lifted the Bible to protect herself," Tavares said. "But he was at her side and shot her six times—in the head and the chest." The farmer hid in the forest, then ran to the meeting and told the others that Sister Dorothy had been killed by Fogoio and Clodoaldo Carlos Batista, known as Eduardo.

One of the farmers rode his motorcycle to Anapu. "You must understand," Tavares told us, "that in Boa Esperansa, there are no phones, no electricity. Her body just lay there. It was raining, so it wasn't until one o'clock that the call came to Altamira."

Tavares received a summons an hour later, and he and Silva arrived in Anapu at four. "The body still hadn't arrived. The police had taken a helicopter from Altamira, but it was raining so hard they had to take a pickup truck to Boa Esperansa to get the body. It sounds inhumane, but it was the only way to transport her."

Tavares said that Marina Silva met the body in Anapu and stayed up all night with the rest of the town. "The minister was very moved and very angry. They all were very angry," he said. "We were angry, too. We had offered to protect her for a long time, but she refused. She said that the people who needed protection were the *posseiros,* not her."

According to Tavares, "the world descended on Altamira" the next day. The government promised to track down the killers; that was Tavares's job. He sneered, "They were tough men when they killed a seventy-three-year-old woman, but they couldn't survive at all in the jungle." First, Fogoio, the triggerman, turned himself in after six days of hiding, and he told the police various stories. First he claimed that the killing had been ordered by Francisco de Assis dos Santos Souza, the president of the local branch of the Rural Workers' Union, and Sister Dorothy's closest colleague. "He was a bad liar," Tavares said, "and we never believed that story." (Two days after the murder, Souza himself received a note saying, "Sister Dorothy has been killed. You are next.")

Then Fogoio claimed that he and Eduardo had split up and that Eduardo had fled in the opposite direction. "We wasted some days with that lie," Tavares said. "Finally, when we told him he would be treated well if he told the truth, he let us know where Eduardo might be. We surrounded the area, and I smelled him before I saw him. He hadn't bathed in over ten days."

The police also arrested a purported intermediary between the rancher Moura and the killers, and finally Moura turned himself in. "The story is that he was going to pay fifty thousand reais to kill Sister Dorothy, and he was going to raise the money from a consortium of ranchers," Tavares said. "But we don't think any money was paid."

During the weeks following Sister Dorothy's death, it appeared that a sustained international effort to shame the government into providing security and greater forest reserves would be mounted, similar to the aftermath of the Mendes murder. But this time the government preempted these efforts by denouncing the murder, sending the army into the area, and creating 8.2 million hectares of reserves. Nationalists decried the attention given to an American woman, when so many Brazilian men had been killed in the same cause. The Catholic Church expressed regret about a violent death but said nothing about the circumstances in which Sister Dorothy had lived and died. Within two weeks of her death, two rural workers and a union leader were assassinated; two ranchers were murdered—all in Pará, all allegedly interconnected.

Once the perpetrators had been apprehended, the story left the headlines, and life returned to normal in the Amazon.

Pope Benedict XVI will no doubt stay on message that the Church should stay out of politics; after all, he was the primary architect behind the Catholic Church's break with liberation theology. He personally reprimanded Leonardo Boff in the early 1980s. The Amazon had served as a magnet for foreign clergy committed to class struggle, but that imprimatur has been lost. Sister Dorothy's death magnified the consequences of religion's straying into politics.

Even political parties, which inherited the Church's advocacy role, have generally proven ineffectual in the Amazon because poor people provide insufficient fuel in any political machine. Often disenfranchised from voting and financially impotent, the landless present few resources for politicians, especially when the system resembles a kleptocracy. During our recent visits, we observed massive corruption investigations that dismantled the state legislatures in Amazonas, Rondônia, and Roirama. The federal senator from Pará attracted constant press scrutiny regarding his private jet and other markers of wealth. Lula's government itself has slowly dissembled, as one major figure after another has resigned in corruption scandals. The poor farmers in Anapu and Boa Esperansa could not afford an entrance ticket into this system. The farmer who hosted Sister Dorothy on her last night told a reporter, "We have lost a great defender. No one else will help us." Democracy was failing them.

Itaituba is a clone of Altamira, separated by 186 miles. It is a city of about one hundred thousand people congregated into a dusty or muddy (depending on the time of the year) collection of wood shacks and white stucco buildings. The city has open sewers and drainage, some paralleling the roads and some emptying into the dirty river. The Tapajos River, so brightly blue near Santarém, is brown at Itaituba from all the dirt and sediment running from the town and the mines nearby. We had been in Itaituba before, during the mining boom of the 1980s, when the airport was so busy that planes took off in opposite directions, often at the same time, ferrying miners and their gold. Now the airport is calmer, but the city still bustles. Itaituba is geographically superior to Altamira, and should be several times Altamira's size in ten years. The BR-163 road from Cuiabá to Santarém, when it's paved by

2007 (*se Deus quiser*—if God wishes), will branch off to Itaituba, making it a major grain port.

Maria Elza Ezequiel de Abreu came to Itaituba in 1981 from the northeast state of Maranhão (a primary source of east to west immigration) with her husband, who was a miner. "Not long after we married, he got malaria," she said. When she heard the news, she dutifully left town to be with him at the mine. "That's when I saw the life of a miner, which I don't think he wanted me to see." She shook her head as if she had been duped.

"That lifestyle is disgusting. It's an illusion. There isn't much gold, and they just tell themselves about how rich they're going to be. No one gets rich. Most of them lose whatever they have on drink and prostitutes while their wives and children starve in the city."

Constrained by the economic and social realities of being a woman in Itaituba in the early 1980s, Elza could only bide her time idly. "My husband would come home from time to time and be a husband and a father, and then he would leave again." She eventually found her calling. "My grandfather was a union delegate," she said, "and always in danger of being imprisoned. One day they told him the police were looking for him. He ran to the forest with his family, including my mother, who was pregnant. They stayed hidden for four days, and returned home at ten o'clock at night, just so I could be born."

Raised by her mother and grandmother, she developed an idealized view of the union movement her grandfather had joined. "I decided what my grandfather didn't live long enough to do, I'll do. Anyway, it felt strange just to stay in the house just hoping that one day my husband would arrive and give me money."

In 1987, she joined the MST (Movement of Landless Workers). A year later, she camped out in protest in front of the local INCRA (National Institute for Colonization and Agrarian Reform) office, the government agency in charge of land distribution and settlements. INCRA's responsibility essentially is to find abandoned private land or unused federal land, have it demarcated, make plans for infrastructure, and develop a settlement plan. She camped out for fifty-two days.

"I would go home at four in the morning and cook meals and then come back."

The protesters eventually received a plot of land three hours by car from Itaituba called project Cristalino I, where they settled. The early days were especially rough. All that existed was degraded forest, which quickly swelled with scores of families living in makeshift houses with black plastic roofs and with no infrastructure or public services. But Elza said this group managed to create a viable community, and she served as its health agent for six years, then as its president.

She eventually moved back to Itaituba to work with the Rural Workers' Union. In January 2003, she became its president. She works out of a boxlike whitewashed building with the colorful sign of the union and its motto: OUR STRENGTH IS OUR ORGANIZATION.

"There are fifteen thousand members in forty-eight communities," she said. Her agenda reads like the menu of concerns everywhere in the Amazon: deliver the social services of health and education to the rural people, so the rural people don't become the urban poor; provide the rural people with documentation of legal ownership of their lands; protect them from *grileiros;* provide them with agricultural know-how so they can support themselves; enforce the laws against illegal logging and drug trafficking; develop programs to empower women; teach rural agronomy and provide rural education on agroforestry.

Elza was one of the few leaders of the Rural Workers' Union with whom we met who hadn't yet been threatened. But Elza had painted a target on her back by taking up this cause. Her counterpart in Rendenção, Valmisoria Morais Costa, told us of the murders of two of her coworkers in the past three years—a local officer nicknamed Dezinho and another named Ezequiel de Moraes Nascimento, both of whom were murdered in front of their wives and children.

The INCRA offices in Itaituba aren't even in Itaituba, and they're not even offices. To get there, we took a barge—big enough for twenty-four cars—that crossed the Tapajos to Mirituba every hour or so. Many of the barge passengers live in rural areas that are hours away by bus (and then hours of walking). These rural folk on the barge clung to fresh supplies from the big city. There are no paved roads in Mirituba, which is particularly problematic in the rainy season because INCRA sits atop a slick muddy hill. Anyone visiting INCRA needs determination and strong legs.

Luis Ivan de Oliveira, fifty-three, has been INCRA's director since the middle of 2003. He was appointed by the Lula administration and once worked with the Rural Workers' Union in Itaituba and with the agricultural workers in Santarém. He and Elza knew each other from their work with the MST.

"Because of the work I did with the MST, I received a death threat," he said. "I took my family and left Itaituba for forty-five days. I went to Brasília to make a formal denunciation of the loggers who threatened me, but"—he laughed—"nothing was done. At least, I think they [the loggers] stopped being angry."

He said that in the past few years "the land occupation situation has greatly deteriorated. I think much of it is because of the paving of BR-163 and the invasion of soy. If precautions aren't taken immediately, the area will become too dangerous and isolated for the rural workers. It will be too late for the state to intercede. The private interests are so powerful."

Behind his desk, neatly stacked in scores of piles, stood the solution and the problem. "Here," he said as he stood and grabbed a pile, "are the applications of dozens of people to own land." He sat with the pile in front of him. "You look at this, and you say, 'So, give them title and give them land.' "

He shook his head. "Not so easy. Many of these applications have the name of *laranjas* [literally "oranges," although in this context it means people who apply for land and later pass it on to those who hired them] from the south, from Curitiba, Paraná, Santa Catarina, Rio Grande do Sul. These *laranjas* are not the real owners. People use their sisters, wives, and girlfriends to get land from the government, and then they sell it to themselves. They sell it for a higher price to cattle ranchers or soy farmers. Most of these women don't know anything about the land, not even where it is."

The applications are too numerous to count, the land too diffuse to catalogue, the tools for demarcation too scarce, and then when all the other pieces fall into place, it turns out the applicants for the land don't even know they've applied. "I've been here for twenty years," Ivan said. "I know the land like the back of my hand, but I don't know how to do this bureaucratic part alone."

In the 1980s, forty INCRA employees worked out of this office. Now there are twelve. His counterpart in Santarém hasn't hired any-one in twenty-one years. Fernandes Martins Ferreira, who works in INCRA's São Feliz office, complained, "We don't have a single topog-rapher. And we cover three municipalities. We're without inspectors. There's a big deficiency here."

Ivan's jurisdiction covers an area the size of Belgium. "We need to visit the areas indicated in the applications to make our rulings. We need to oversee the areas with large land titles so we can prevent more armed conflicts. But the *grileiros* have many more resources than INCRA. And IBAMA in Itaituba, which we need to help the enforce-ment, doesn't even have a director. Two thousand families have been removed from the Gleba Leite locality, and the valuable wood has been extracted. People are dying in these conflicts. The government is aban-doning these rural workers. Without assistance, they leave and come to the city. It's just more misery and destitution. But we can't abandon them."

The MST is as close to a self-help organization as exists that is dedicated to invading and occupying private land. The MST takes over unused or unproductive land by having scores of families invade the area. The MST believes it is creating efficiencies in the Brazilian economy, be-cause an unproductive resource is being expropriated to provide suste-nance. Others consider the MST thieves.

Charles Trocate, one of the coordinators of the Maraba chapter of

the MST, embraces the view that underused land should be reallo-
cated. The son of a Krikati Indian mother and a Paraense father of
African descent, he has been a rebel from the age of five, when a
schoolmate yelled at him, "Indians eat people! Indians eat people!"
and Trocate stuck a pencil in the bully's ear. He's short and lean, and
his left eye has a motor problem that causes it to squint shut when he
smiles. He stopped going to school in the first grade. By dint of his
heritage and environment, he faced insurmountable odds of becoming little more than a beggar.

Yet he is hardly one of the forgotten people of this country. His sister taught him to read, and he became enamored of the poetry of the
Uruguayan Mario Benedetti. He reads Spanish, too. He speaks in
complete paragraphs and has written two books of poems. His favorite, *Rise Up Everyone!,* ends with his personal credo: "These are the
years to live! These are the years to overcome!"

The MST arose from the pools of *posseiros'* blood that stained the
land in the late 1970s and early 1980s, Trocate explained. "These were
groups of men who would get together and invade a farm or area of
land with guns and machetes. They would cut the land into lots and
each family would protect its own through armed combat, if necessary.
There were many deaths and murders at that time. It was a time of individual and unequal conflict."

The *posseiros* inevitably lost. Without the means of acquiring title or
superior firepower, they moved on when death threats became unbearable. "The MST philosophy of occupation is different," Trocate said.
"Instead of groups of armed men, whole families participate in a
peaceful takeover and live on the land. Central organization allows all
of the families to work together for the same goal instead of the precarious individual defense of small lots against powerful farmers and
their *pistoleiros.* After a successful occupation, the MST stabilizes the
area by planting essential infrastructure and training and educating
workers and children."

Trocate sees his aim as organizing the grassroots level to correct an
oversight of Brazilian history. "We never had agrarian reform," he

said. "We are an interrupted nation. We never had a phase of agrarian reform before industrialization, and we never reformed our capitalism to democratize our land and our economy. Through a form of collective resistance instead of individual conflict, we will do that."

With Brazilian pop music playing as background on his computer, he mapped out a four-point plan to justify these takeovers: "First, we need a community of people, teachers, guitarists, pilots. This mix gives us the opportunity for cooperation and strength—different than if families lived individually. Second, any occupation needs to be large. Numbers of people provide more security than a few guns. Third, we have multiple leaders who act at the same time, coordinating their work. The MST doesn't use a pyramid hierarchy. Our leaders, when recognized, are subject to threats and violence. Fusquinha and Doutor, two of our most vocal leaders, were murdered in 1998. Another, Ivo Laurindo do Carmo, was brutally killed in 2002. Fourth, we follow several different plans of action to achieve our goals of education and agrarian reform. We form assemblies, organize fairs, and integrate our goals into local politics to reach many different segments of the community."

We were nonplussed. His "radical" plan defines every small town in America. Except that the U.S. government gave many of the farmers the land if they agreed to live there, a bargain Trocate believes the Brazilian government owes its citizens. He also wouldn't oppose the government's compensating a landowner whose abandoned land is occupied by landless families. He believes that's an acceptable retort to those who charge the MST with theft. His concept is similar to eminent domain, except land goes from private to private hands, with the government acting as the transactor.

The problem, too, can be even more complicated, because not all land is privately owned. In many cases, he said, private owners claim title over public land. In the late 1990s, INCRA launched a "prove it or lose it" campaign, which required all landowners in the Amazon with more than ten thousand hectares to prove the validity of their titles. According to Raul Jungman, the head of INCRA at the time, sixty-one million hectares reverted to the state when legitimate title could not be proven.

The MST in turn also invades land owned by the government. The government resists these efforts because of the high costs of providing an infrastructure to support such a community. "The multinational companies, the large landowners, the state government," in Trocate's view, have produced nothing but slavery, deforestation, and pervasive poverty. And he sees Lula, once revered by the left, as an obstacle to any fair solution. "The Lula administration is not going to have the vision to increase employment in Brazil. If the land regulations aren't radically changed, the landless people's situation won't improve at all. Lula came in and imagined that he'd be able to make a negotiated transition of power as opposed to a change through conflict. But the same sectors are still being favored. Nothing substantive has changed."

Nothing substantive has changed? It was fruitless to try to convince this young man, fueled by the fire of social activism, that change takes time and that it doesn't always move in a linear direction. That Lula could become president. That an Internet would serve as a means of organization. That cell phones would serve as sentries to warn those in danger. That solitary *posseiros* would give way to collective settlements. That Charles Trocate, poor and illiterate, would become a multilingual poet.

INCRA's compound in Maraba covers several hectares of land within the city limits, and its headquarters are new, air-conditioned, and have beautiful wood adornments and lots of glass. If the government had wanted to make a statement that agrarian reform around Maraba mattered, this compound is doing so. At least architecturally.

Inside the building, functionaries sit at desks in partitioned interview cubicles, taking down information from bedraggled men who spend hours waiting to be heard. Many of the voices speak unschooled Portuguese, mentioning death threats or health problems, or they have questions about when promises made will be met. The INCRA employees are pleasant and sympathetic, if not at all effective.

The expansive grounds of INCRA serve as a makeshift home to more than two thousand people, who have come to petition their government for land. Squalid shelters with black plastic roofs, held up by wobbly planks, infest the usable space on INCRA's property. It is a refugee camp, a Hooverville. Hungry babies complain of their pain, rapid voices barter what they have too much of for what they need, and a sour smoke hangs in the air over the muddy ground, combining with the smells of latrines, burning refuse, and dinner. In the shadow of INCRA's edifice, this staging area has spread, reminding the government workers inside of their unfinished mandate. No one can come from or go to work without visualizing what their work entails.

Arlindo Alves recently arrived in the camp. He made his place near José Artur da Silva, already there for five months waiting to hear about land grant opportunities. "I don't know where the farms are that we're trying to get to," he said. "People have told me but the names just fly out of my head. I don't care. I just want land. We need land to work so that we can survive," Alves told us. Born to tenant farmers in the state of Maranhão, he came to Pará to work as a farmhand. "I spent seven years living on the farm of my ex-boss. He said when we started that he was going to give us some land to plant our own food. He made a lot of promises, but when the time came to pay up, he failed. He didn't keep them. So I left the farm."

Alves moved to Maraba with his wife and five children, but couldn't find work. A farmer lost in the city, he joined what he called the "movement," an organized effort to wrest land from the government. Having heard that INCRA was going to legalize a settlement outside of Maraba, he decided to camp out until that happened. "It's better to be on the farm than in the city," he said. "In the field, my kids see their father working and learn good values. Here in the city, they see their friends pass by on the street on their way to go play or commit some crime."

So when will he be able to move from here?

"So Deus sabe." Only God knows. "I need to support my children. To do that, I need a piece of land to work. I want a more secure life and one without hunger. I often go to bed hungry now."

The nearest INCRA settlement to Maraba, the type of place these people dream of calling home, is the Primeiro de Marso. Begun seven years ago, it's now abandoned. Once INCRA certified it, no one but the original settlers could occupy it. For a reason no one could explain to us, the original settlers had given up and left. Four structures line the entry road: a half-finished concrete block building with no roof and rampant weeds, a shuttered brick house, a small empty shack, and a bar. It was a scene of irony in light of the hopeful visions of so many squatters in the nearby INCRA headquarters.

An even more disheveled collection of shacks lies about a half hour away. Our driver told us it was an "illegal," meaning that it hadn't been approved by INCRA. It's best to approach the "illegals" on foot, preferably with a female in the lead to minimize the chances of being shot.

INCRA had already declared this farm, Fazenda Lundi, to be "unproductive," so the bureaucratic process was supposedly in progress. Most of the area was overgrown with wild grasses and shrubs that sprouted indiscriminately.

Horacio Rodrigues Chaves, sixty-three, arrived with his family on October 17, 2003, with seventy other families. "We've been here every single day for eight months," he said, "and still INCRA has not started the process to make it legal." In the "legals," INCRA assists the new residents with baskets of basic foods. Here they live off leftovers from other settlements. "We put up the shelters in the first two weeks we were here," Chaves explained. "We did not expect to wait eight months and still have no idea when we will have a legal settlement, but eight months is not long to have a camp."

Yet he was frustrated. "We're here with our mouths open. We're waiting for someone to help us because we've done as much as we can without breaking the law. There's only one cowboy on the farm. There isn't anyone to farm the land. It shouldn't be hard to help us."

José Leandro da Silva, also sixty-three, joined the conversation. It turns out he didn't even live on Fazenda Lundi. "My son had work in another area," he explained. "And I came to be his substitute. I live in Maraba, and when my son leaves, I have to come here, or I will lose his place."

A young girl came by holding a small baby. "Can you watch our house while I go out?" she asked. "My parents left, and I need to go." She shifted her weight from one foot to the other, swaying playfully.

We asked if she went to school.

She giggled. "I've never been to school. I'm fifteen. I have a baby! I don't go to school." She shook her head. Her breasts started leaking milk. She quickly crossed her arms over her chest but stood her ground.

Chaves kidded her. "Why don't you just lock the door and throw the key in the woods? Then you won't need anyone to watch the house." She giggled, and just stared at him, waiting for an answer. Finally, he told her he'd look after the baby, and she ran off.

We took a detour to a holy shrine of the landless movement. Thirty-five hundred families had come to the south of Pará in March 1996 to occupy an abandoned ranch. They received assurances from INCRA that they would eventually receive titles to the land, and food and medical supplies while they waited. Delays ensued. After suffering thirty days of bureaucratic inertia, approximately fifteen hundred squatters embarked on a walk to the state capital of Belém, five hundred miles away, to draw attention to their plight. After three days, they changed tactics and decided to blockade the state highway, insisting that the government supply food and water and buses to transport them to Belém.

They were poor, hungry, looking for land, and willing to stop traffic to make a statement. They presented no clear and present danger; no town was being invaded and no hostages taken. But the state military police reacted brutally at about four P.M. on April 17, 1996, where the road from Maraba to Xinguara turns like an "S," just before the village of Eldorado dos Carajas. The military police killed nineteen unarmed civilians and seriously injured eighty-four others, three of whom sub-sequently died. Nineteen naked tree trunks now stand at the site, shaved brown sticks that look like a small farm of telephone poles. A plaque lists the names of the victims, although their act of heroism is ambiguous. They were victims of state-sponsored crime more than they were martyrs to a cause. They had nowhere to go and nowhere to live, and they'd gathered to protest that fact. They inconvenienced traf-fic. On June 13, 2002, a Belém jury acquitted 124 of the soldiers of any crime, accepting an argument of self-defense. Two commanders were convicted, but these verdicts are still on appeal and may not be decided for ten years.

The senselessness of this massacre lingers on the frontier, almost as much as Mendes's murder. These people died in the name of what? The Primeiro de Marso settlement? The abandoned clearing we had visited? The Fazenda Lundi, where people lived in a perpetual bureau-cratic limbo? What had made them enemies of the state?

The massacre occurred during the enlightened administration of President Fernando Henrique Cardoso, who promptly condemned it. Amnesty International characterized the incident and the follow-up state prosecution as egregious violations of human rights. Actually, these killings were the consequences of the country's repeated failure to address its citizens' property rights. The federal government fostered optimism in the squatters, then allowed that optimism to evaporate. That the state militia—tired, frustrated, scared, uncomfortable—acted like hired *pistoleiros* should have surprised no one. Had they not killed the squatters, some vigilante group may have come along, or the squatters may have died from starvation—nothing good was going to come of this standoff once it became clear that property rights were not going to be granted.

The unwanted costs of the violence on the frontier—including the international humiliation from this act and its aftermath, the costs of investigation and prosecution, the hasty remedial steps taken to assure that the remaining families settled peacefully—stain the land throughout the Amazon. In 1996, the Brazilian government was not at war with its citizens; yet its inability to protect them, to provide them with the requisites of human dignity, brought about the same result.

At a place like Eldorado dos Carajas, the nexus between the human and natural destruction was apparent. No one would have died there if the government had had the resources to: survey the land, make determinations as to what had been abandoned and what was unproductive, offer eminent domain payments to lawful owners, or void forged titles. If the government had then been able to redistribute land, either for free or through a credit system, the result would have provided a widespread benefit, not only in the obvious affirming of human life but in the resulting positive economic contributions these squatters might have made. What happened at Eldorado dos Carajas is that the paintings in the Louvre got burned, as the diplomat Everton Vargas had warned.

The government lacked the resources not only to provide a solution but also to prevent human nature, as it appears on the frontier, from taking hold. Detractors are missing the point when they condemn the government for possibly lacking the political will to intervene. Political will itself is a consequence of the availability of resources. Funding the purchases of land for the landless in the Amazon would have diverted money from programs in the more populous urban areas. How is a government supposed to protect a supposedly international resource if the constituency to which it belongs is quick to contribute criticism and slow to contribute solutions?

Here the burden on the country of the Janus-like blessing of the Amazon came into focus; the promise of a "land with no men for men with no land" became a pool of blood. The condemnations from international organizations about Brazil's failure to protect human rights, as well as the related criticism for the country's stewardship of the environment, were of marginal benefit. The violence at Eldorado dos

Carajas, as elsewhere, was a tragic symptom of the problem, not the problem itself.

Our e-mail exchange with the son of Samuel Benchimol, who had as concise an understanding of the interconnectedness of the issues, concluded that what was at stake in the Amazon was a quest for "national sovereignty." By all measures, the Cardoso administration understood that national sovereignty only succeeds with national participation. But still the administration lacked dependable institutions and available resources to address decades of exclusion. Cardoso himself acknowledged this in his autobiography, in which he wrote, "I knew that the only effective way to avoid further bloodshed was to try and solve the problem at the root of the dispute. But doing so would mean addressing issues that had festered in Brazil for five hundred years." The international hue and cry, reminding Brazil what it already knew—that human rights need to be protected—may have been cathartic, but it was as ephemeral as the smoke that covers this land when the dry season comes.

The March 26th Settlement is named for the date in 1998 when Fusquinha and Doutor, the MST leaders Charles Trocate mentioned, were murdered. The settlement is located about twenty miles outside of Maraba, and there is a large mural painted on one of the long one-story buildings at the top of a hill. The only access from the main road is a path—bumpy, muddy, and wide enough for an oxcart. The other building also has a mural: ESCOLA CARLOS MARIGHELLA and a picture of a dove on the school wall. When we arrived, a group of men leaned against the wall, taking a break.

"You'll want to talk to Sebastião," one of the men said, fending off an interview. "But he's busy now. They're having a meeting." That was a brush-off. There was no such meeting, that was clear. "Where is he?" we asked. The men looked around, debating with their eyes. Finally, one of them went into the school and brought out a man claiming to be Sebastião Araújo. He was busy, he said. There was a community meeting, no outsiders allowed. Better to talk the next day, he said. It was difficult to say whether he was being unfriendly or wary.

The next day a different Sebastião, open, warm, and friendly, led us on a tour of the settlement. Though the settlement was still waiting for final approval from INCRA, 274 families were living there. All of the children were in school, 480 of them. "We have enough teachers, and they are trained well," Sebastião said. "But we don't have a health center, so when people are hurt or sick, they have to go by bus to Maraba."

Sebastião said that the settlement began with an invasion of a ranch (Fazenda Peruana) five years ago. The settlers quickly built improvised shacks and slowly added infrastructure. "We built everything with our own hands," he emphasized. "No government help. And we don't even know if the government will let us stay. INCRA may give us a parcel of land somewhere else."

Here at the March 26th Settlement, it looks as if this invasion has resulted in something permanent, more so as we walked with Sebastião deeper into the area. The owner of the property hasn't threatened them, and they've been careful not to create any conflicts with the cowboys, Sebastião said. "But we won't know whether any of this land is ours until INCRA releases its decision. Until then, our only option is to keep working the land, organizing, and praying."

We asked him how difficult it was to make plans during five years in limbo, knowing that the government could change its mind and send the police instead of an envelope of land titles. "Difficult," he said, and shrugged.

Besides the school and a rice storage building, there are three churches, a community building with a TV, several small general stores, and, of course, a soccer field. It looks very functional. "Unfortunately, sometimes we have internal problems," he said. "You can never have one hundred percent, and there are a few abandoned houses here. But the great majority are still here working." The settlement appears clean for such a densely occupied place, although there is no trash pickup or street cleaning. The general stores display their wares neatly, batteries and candy hanging side by side with onions and fresh meat.

The community has twenty coordinators, one per sector or block. The blocks, originally divided into family groups, expanded into groups of five to ten families. The settlement has three hundred official members of the MST, which means that some families live there without registering or participating in MST activities. "It's not like we can kick them out, though," Sebastião said. "They're all part of the fight." There also are members who don't live in the settlement, and whose job is to petition INCRA to act. "If we don't complain and demand our rights, nothing will happen," he explained.

Each wood-and-palm-frond house seemed to have at least one MST poster on it as well as signs such as PEACE, CHE GUEVARA, or CONGRATU-LATIONS MARCH 26. As we walked through the settlement, children gawked at us, giggling and pointing. Adults greeted Sebastião warmly.

Sebastião lives in a modest house in Group 5 with three rooms: a bedroom, a kitchen, and a spare room hidden behind a curtain. The backyard has an outhouse and various fruit trees. The house is built in the shade of the trees and has some circulation, so the stifling heat of the Amazonian indoors isn't intolerable.

Sebastião and his wife, Dona Maria, participated in the live-in outside the INCRA headquarters in Maraba while they waited for land. "I didn't like it there, you can see," she said. "There's no bathroom. They have to cook outside in the open air. Here it is much better."

She and her husband decided to go back to school to improve their reading and learn computer skills. She's in fourth grade; he's in fifth. The school, she said, planned to get two computers for the students, so the community could learn how to use them.

As Sebastião walked past the soccer field on his way back to the school building, he said, "This is not a hunt for power. It's a quest for dignity and independence." That much was clear. Why else would a grown man sit in a fifth-grade classroom every day to learn how to read and work a computer, while his wife labored in fourth grade? Why else would they have created a communications and transportation network to make sure that they were able to take the sick among them to Maraba before they died in the settlement, waiting for medical help to arrive?

Sebastião derided the notion that this settlement was succeeding because of superior location or any other physical advantage. "We succeed because of who we are," he said. "In the end, what we produce depends on how much time people spend and the will they have to work. For the thing to work, you have to involve the youth. If you don't give them a part, they become discontent and go to the city." Sebastião said that if the settlement stands for any principle, it is education. "This is the fight," he said, pointing toward the school. "We need to have people educated. I'm learning. We're seen as good by some people and disliked by others. But I'm doing this so that one day my children will see what I created and say, 'This was my father's struggle!' "

CHAPTER 17

A WAY TO SAVE THE AMAZON

—

For much of the twentieth century, Manaus, a speck of settlement surrounded by an unnavigable green sea, has been the capital of the Amazon jungle. Landlocked and lying along the confluence of the brown Amazon and the black Rio Negro, Manaus is nothing more than an oasis of two million people cut off from the rest of the world, as disconnected from Brazil as Tasmania is from Australia. Most people born in Manaus never venture far, if they leave at all.

Few roads access the city. One goes north to the state of Roraima and on to Venezuela. It is mostly for adventurers and cargo haulers and bears no real commuter traffic. There's a road to the waterfalls in Presidente Figueiredo about two hours away, a local tourist destination, and there's a road heading east to the soy and fertilizer ports in Itacoatiara, three hours away. A paved road once ran from the south bank of the river five hundred miles to Porto Velho, but it washed out in the early 1980s. It is now a rutted jungle path that is impassable during most of the year, although rumors of its reconstruction abound, delighting the Manaus business community and frightening the world's environmentalists.

Airplanes serve the city, but flights have been cut drastically due to the financial problems of Brazil's airline industry. An Air France 747 used to stop there on its way from Lima to Paris, and Varig had a daily flight to Miami. No link to Europe exists now, and the only connection to the United States is at 1:40 AM on a Bolivian airline flight stopover. The other option is to fly four hours south to São Paulo in order to fly ten hours north.

Anyone in Manaus who has been outside the city has most likely traveled by boat. Landing stations sprinkle the river's edge, the busiest being downtown alongside the malodorous fish market, where ornate ironwork intended to replicate Les Halles in Paris serves as a relic of the rich rubber era. The downtown passenger port is nothing more than a jumble of vessels huddling together, each sporting a hand-painted wooden plank announcing a destination and time of departure. The smells can be overwhelming. Men piss where they please. Garbage accumulated during multiday river travails gets dumped into the water or dragged across the shoreline and dropped in open containers. Deck hands and pilots of small boats, killing time waiting for passengers, play cards and drink raw cane alcohol. Laborers, bent double under cases of soda and beer and sacks of foodstuffs, move steadily back and forth from truck to boat. Prospective passengers balance themselves on the planks connecting the shore to the boats and the boats to each other, asking if space is available, about the price of passage, and how long it will take to get there.

The boats vary in size from the four-layer-wedding-cake hulks that hold about 300 passengers—30 in cabins and 270 in hammocks—to the motorized dugouts covered with roofs of dried leaves and enclosed by mosquito nets, sweltering warrens of twenty or so hammocks each. Some of these river journeys can take up to two weeks. Santarém to Manaus is less than an hour by plane, but four days on the river. Tabatinga is almost two weeks away on many of these small vessels. The boredom of a fortnight journey on the Amazon and its tributaries is indescribable. There's nothing to do, and after the first few hours, the scenery never varies. Life approaches the monotony of the purring motor. The days consist of awful heat and pelting rains, and the nights can be cold and scary because of the turbulent currents, dangerous flotsam, and throngs of bugs. Toilet and dining facilities do not exist on most of these boats. For those arriving in Manaus for the first time, the travel experience resembles the rigors of the arrival of immigrants to the United States from Europe.

In its history, Manaus has had two moments of prosperity, one a hundred years ago during the rubber boom, when it controlled the world's supply; the other is now. With nearly two million people living in the capital of the state of Amazonas today, the city has flooded into the surrounding forest and suffocated the plant life with prefabricated housing developments, traffic circles, and shopping malls. The bus routes work in a hub-and-spoke system, and the central stations teem with commuters and enterprising kiosks. High-rise office buildings and residences have sprouted everywhere in the skyline, and the Ponte Negra district along the Rio Negro resembles Collins Avenue in Miami Beach, with its lineup of luxury apartments and its string of boardwalk open-air bars along a boulevard street. Gated communities of single-family homes line many of the major arteries. Traffic at rush hour can be brutal, especially during the rainy season, when the roads flood spontaneously. Pockets of urban misery peek through the developments here, as in every large city in Brazil, although Manaus's isolation has so far spared it the plague of drugs and urban violence seen in the south. Open sewage and litter plague widespread areas of the city, but the pace of public construction suggests that the filth is being cleaned up. With a heavy military presence—Manaus is the headquarters of the armed forces in the north—and reliable employment in the successful free-trade industrial zone, a middle class has emerged in Manaus, more so than in any other city in the Amazon. Few reminders of the rain forest linger in this metropolis.

Due to the breadth and preservation of the rain forest in the state of Amazonas, where a stunning 98 percent remains untouched, its governor is one of the more influential people controlling the preservation of what the world believes "the Amazon jungle" to be. Amazonas makes up 60 percent of Brazil's Amazon. It's the size of Acre, Mato Grosso, Rondônia, and Roraima combined. Carlos Marx, the satellite map reader who concluded "no one can get there," considered it the Amazon's safety net.

The power of Brazilian state governors in their federalist system exceeds that of their counterparts in the United States—perhaps not on paper, but certainly in practice. Democracy has diminished but not erased the Latin American tradition of a *caudilho,* or strongman. When Blairo Maggi decided that the state of Mato Grosso would become soy-friendly, it did. When Jorge Viana envisioned the state of Acre as a model for "green development," that's what happened. Amazonas's governor Eduardo Braga wants to provide economic and social incentives to his constituents while comforting the world's environmentalists.

In the meantime, he only wishes he were up to the task of keeping his appointments on time. "You know who's out in the waiting room, don't you?" Governor Braga asks on a makeup visit—once, we waited three hours, only to be told, along with a dozen others, that the governor was "terribly sorry that he had run out of time." In flawless English, Braga pointed out that we had line-jumped mayors from Boca do Acre, Labrea, and Coari. "You know how long it takes them to get here? Days. And now they've been waiting for more than four hours. Maybe more. This doesn't make me happy. But I don't know how to do better. Everyone has something they want to discuss. And there's only one of me." He confessed, "I think I'm a much better administrator of the state than I am of my own office. At least I hope so."

Braga, in his mid-forties, is trim and well barbered with an instantly engaging manner and a raspy voice from nonstop politicking. In between sentences he slumps, revealing the wear and tear of his job, the governance of territory as vast as the land between Chicago and Juneau. "I am constantly tired," he confides. "There is so much to do. So much space to cover. And so much of it is empty, which presents different problems."

Braga's state job is made much easier by at least two serendipities that he inherited. The first could be found in the Zona Franca, the heavily subsidized free-trade zone, where dozens of foreign firms have assembly plants. "It changes everything for us. It gives us flexibility and resources that the other states don't have."

His bailiwick also includes the Urucu gas and oil fields, the second windfall. "Gas will give us the energy to allow industry to grow in Manaus. Gas will give us the energy in the small towns to improve their quality of life. Gas will give us the money to do other things, to improve social services here and to have programs to develop the rest of the state in a way that protects the environment."

What once was the rubber capital of the world has become the industrialized center of the rain forest, and that explains, in large measure, why the deforestation rates in the state of Amazonas have been remarkably low. The initial plan for Manaus was simple and straightforward and had no agenda for the environment. The city was made a duty-free port in 1957 to attract consumers from the rest of the country, and to encourage them to make large purchases and, while shopping, to stay for a tour of the exotic Amazon. The plan justified an aviation connection between Manaus and the rest of the country, a modest start to its integration.

With the arrival of the military dictatorship came central planning and increased concern about the Amazon's emptiness and the isolation of Manaus. The military desired more aggressive steps toward annexation. The idea in 1967 was to make Manaus a comprehensive free-trade zone, not just a shopping haven. In the initial plan, Manaus received broad tax and duty concessions for thirty years in order to attract domestic and foreign investment. The government intended that the long-term concessions would attract labor-intensive industries and allow a reasonable time for recoupment of initial capital investments. Import duty exemptions were given to some finished items imported into Manaus and, more important, to semi-finished items or component parts. The latter exemptions gave rise to an industrial park of assembly plants. In turn, the finished products received exemptions from tax on the value added in Manaus. There were additional incentives in the form of income and sales tax relief and favorable conditions for the purchase and improvement of real estate.

The specific programs and the amount of concessions have varied from time to time, but the underlying concept of creating favorable conditions for import, assembly, and subsequent export have remained in place. Manaus, in turn, has flourished. At the outset, tourism swelled, and the first-rate Hotel Tropical arose as an anomaly of luxury in the erstwhile forest along the banks of the Rio Negro. (Because of the pH level of its water, the Rio Negro provides a relatively bug-free environment.) More important, the Zona Franca, as the economic zone is called, drew the interest of hundreds of major multinational corporations such as Honda, Nokia, Minolta, Gillette, and Harley-Davidson. The area sports one factory after another, hidden behind whitewashed walls and fences. Some of the companies provide soccer fields and swimming pools on-site. These are towns within a city.

The oil shocks of the 1970s and the general malaise of the Brazilian economy throughout the latter half of the century, as well as political tinkering with incentives, contributed to dips in this artificial economy. The general opening of the Brazilian economy to foreign goods in the 1990s also contributed to these economic low points. But the national government's commitment to Manaus's economy and the regional support for the concept haven't wavered. Businesses in the rest of the country have griped about the advantages Manaus receives, but the program now provides its own best protection. Any challenge to the favored-site status would jeopardize billions of dollars in sunken investment and throw hundreds of thousands of people out of work. More than one hundred thousand people are directly employed in the Zona Franca, and all sorts of service and support industries depend on those one hundred thousand jobs. The total of all sales from the economic zone came to $14 billion in 2005. More than $2 billion of exports left the Zona Franca in 2005, including more than $650 million of cell phones, more than $150 million of motorcycles, and $75 million of color televisions. Recently, the fiscal incentives have been extended to 2023 in order to promote further long-term investments and to allow investors time to recoup their capital.

The Zona Franca has been as successful as any other important environmental initiative in the history of the world, though it was never intended to be. It is inconceivable that any of the generals and economists who conjured up the plans cared about global warming or preservation of biodiversity; these concepts didn't even exist in 1967. However, by providing a profligate source of employment, the Zona Franca has attracted tens of thousands of people from the rain forest and kept tens of thousands of people in the city—all of whom might have had to resort to deforesting for survival.

The Zona Franca answers the question "How do you stop deforestation?" Give people decent jobs. If you give them jobs, they can afford houses; give them houses and their family has security; give them security and their vision shifts to the future. In Manaus, it's unlikely that women will lose five of eight children before the children are two, as happened to Dona Maria Raimunda in Altamira. And it's unlikely that children in Manaus will suffer the tragic adolescence of Marina Silva. It's more likely that these children will receive an education.

These benefits arise as the unintended consequences of a program meant to integrate the economy of Manaus into that of Brazil. Nonetheless, the example provides a lesson in comparing schemes of negative enforcement with those that reward good behavior. Regulations, fines, and other forms of punishment, which make bad behavior costly, have superficial appeal but dubious effectiveness in a region where civil and criminal law structures haven't matured. Laws that are on the books but that cannot be enforced rarely deter the behavior they're intended to address. On the other hand, the Zona Franca and its incentives, which promote behavior that benefits the environment and that trigger widespread positive repercussions, provide a useful model for enforcing laws by rewarding good behavior. Tax credits and government subsidies favoring productive enterprises are other examples of providing rewards.

However, a solution as simple as subsidizing employment in the Zona Franca highlights the complexity of deciding how to distribute resources. The advantages the Zona Franca enjoys are expensive to Brazil in that they decrease revenues from value-added taxes, import and export duties, and so on. While these costs are offset by the economic and social benefits the Zona Franca generates, they still represent a high price for sponsoring effective environmentalism. The rest of Brazilian business looks jealously upon this program, knowing that if the program were instituted in another part of the country, it would probably yield just as much employment and quality-of-life benefits.

The costs of the program are borne by Brazil as a whole, absolutely in terms of the cost of the benefits, and indirectly through the reallocation of resources to the Zona Franca from other areas, and, yet, the advantages are shared by the entire world. Living in the midst of this economic marvel prompted Samuel Benchimol to observe that by preserving the rain forest, his neighbors were providing an environmental benefit for free to people far away. He wondered if this was fair.

When he took office in early 1999, Braga knew that in addition to preserving the benefits of the Zona Franca, he had to devise a development program to address the needs of the rest of his state. He also knew that any program he implemented would be subject to controversy, given his role as the de facto curator of the rain forest. Alienating environmentalists, Braga expected, would impede his plans for economic growth because they could mount court challenges and appeal to Marina Silva, the Minister of the Environment, or the international NGOs. Yet, to succeed he knew he needed the support of the business community to create jobs and to provide a tax base to finance public-works improvements. As Jorge Viana said in Acre, Braga knew that "people need to like where they live," the gospel that guides incumbent politicians.

To modulate his standing between the two constituencies, he chose a partner, Virgilio Viana, as the state's Secretary of the Environment

and used Viana as an intermediary with NGOs and international organizations. Viana, a Harvard-trained forester who protested alongside
Chico Mendes, has the credibility to sell Braga's programs to likely
critics. He cut his teeth in state government in Acre, where he served
as an adviser to Jorge Viana (no relation) in developing the concept of
preservation of state forests as a sine qua non to allowing road building. (He's also married to Etel Carmona, the well-known furniture de-
signer who owns the factory in Xapuri.)

Viana and Braga face a natural conflict that they've managed to reconcile by recognizing that neither politician will be effective unless the
other succeeds. Development puts food on the table and money in the
pocket. Connecting the faraway towns to Manaus by roads, telecom-
munications, and a pipeline of patronage jobs provides sustenance for
Braga's tenure. Viana knows that the international community, as well
as Brazil's environmental movement, won't tolerate the deforesting of
Amazonas, and it's axiomatic that roads in the Amazon are the precursor to logging, cattle, and large-scale agriculture.

"I am aware of this tension," Braga concedes. "But we have done a
good job of preservation to date, and you have to understand that sometimes you have to develop in order to preserve. I want people to understand our dilemma and understand that there is no easy solution."

When Braga looks at the map of his state, he sees pressure building from the neighboring states of Rondônia and Acre, and believes
that the population spillover will eat its way up from the south. "I cannot resist this pressure," he says, pointing to the southern border of
Amazonas and towns such as Boca do Acre and Humaitá. "Already,
people are saying that we should give Boca do Acre to the state of Acre,
because it's too far from here and we can't govern it. Soy is coming.
We know that. And when soy comes, the forest disappears. Do we sit
here in Manaus and say it's too far away and there's nothing we can do
about it?"

He plans to create a network of family farms and supporting towns
to provide a bulwark against uncontrolled development. "It's inevitable
that people are going to invade these areas," he says. "It's like near the

Colombian border where there are drugs. No one would suggest that we walk away and leave it empty. No, we need to invest in the economy there for social reasons and for security. If we walk away from that region, the FARC will take over."

Drawing his finger over a map on the conference table in his spacious office, he outlines the roads he needs. "The same idea is true in the south, but we're not fighting drugs. We're fighting loggers and cattle interests. We need to connect Labrea to Apui on the TransAmazon Highway. We need to connect Rio Branco to Boca do Acre. And we need to repave Manaus to Porto Velho, although that's not as much a priority now."

Braga sees two choices for Amazonas on its southern flank: spillover development and the resulting anarchy and violence endemic in the south of Pará, or some semblance of civil society. "If we have roads, we can put IBAMA [the environmental protection agency] there. We can put government agencies there. We can put schools there. We can put health centers there. We can create conditions for family farms that are clearly demarcated and where people can make a living. You think that no controls means no people? No controls means that people just invade the land and do what they want. The people already are there, and we can't leave them behind like a bag of trash. We need to connect them."

But Braga also knows that the cycle of development, once started, cannot be stopped. It is based on an economic, not ecological, choice. "I understand that the small farms eventually will sell out to the big farms, and then you end up with major agricultural interests and small people in search of land. Using the land always brings more profit than leaving it alone. Using a lot of land brings more profit than using a little. Those are rational decisions which lead to a cycle of deforestation. As long as using the land brings more material benefits to people than not using the land, we don't really have much chance. I hope to break the cycle."

He shook his head in frustration. "I wish there were a model for this. But there's no model. We're like a cat going after a mouse, chasing the mouse here and there, and all the time the big dog is coming. The big dog is uncontrolled deforestation. We have to have a plan that's bigger than just one particular problem."

Braga and Virgilio Viana call the program they have developed the Zona Franca Verde, or the Green Free-Trade Zone. "It's a plan to promote sustainable projects that provide alternative uses of the land than deforestation," the governor explains. The specifics vary from area to area. The Zona Franca Verde includes plans "to create product zones, areas of the forest specializing in a particular commodity," says Braga. "We have tried to organize the gatherers of guarana to create price stability. We are having great success in increasing jute production, and we've financed a factory to make finished jute products. We are trying to develop fish farms for exotic fish and food fish. These all are sustainable projects with the goal of keeping people in place. To create sustainable development is to allow rural poverty not to happen."

Braga and Viana also have brought the concept of certified forests to Amazonas, where trees are cut in rotation to assure species preservation and to avoid thinning out areas past the point of replenishment. The work to create the framework for such a forest is daunting. In order to create a certified forest, the foresters must first take inventory of the trees: the land has to be demarcated and each tree needs to be identified and put on a map of the area. Then decisions must be made about the order of harvest rotation, the direction in which the trees should fall, and the route the foresters should take out of the forest. All of this information needs to be stored. It takes years just to set up the structure.

Besides promoting these indigenous industries, Braga wants to see foreign investment brought to the site, rather than the traditional third world export of raw materials to advanced factories elsewhere. "People want to save the forest? They want to help?" Braga asks. "We need resources to establish these programs. Maybe Home Depot wants to build a factory here and will buy only certified wood. Let us add value here. Then we can take those profits and return them to the people."

The area of Carajas has the world's greatest iron ore reserves, but there are no steel mills on-site. Trombetas has one of the world's greatest bauxite mines, but there are no aluminum mills in proximity. Other than Etel Carmona's specialty factory in Xapuri, no furniture makers or cabinetmakers have arrived. "We could export tables, doors, floors, window frames," says Braga. Drug companies constantly scour the region for plants that will provide them with the next major pharmaceutical breakthrough, but there are no production facilities in the Amazon. Nor is there any inclination on the part of any of these companies to acknowledge the intellectual property and patent rights they may have appropriated from Brazil's forest floor. Recently, the pharmaceutical company Squibb discovered that the venom of the Amazon viper *jararaca* worked well as a blood-pressure medicine. This venom became the base for captopril, which at its peak was the largest-selling product for Bristol-Myers Squibb, grossing $1.6 billion in 1991. None of these profits were returned to the Amazon. Bio-piracy stings here, and England's 1876 theft of rubber seeds ranks as the Crime of Eternity.

"It's frustrating," Braga says. "It's frustrating when the Kyoto Protocol does nothing to help us. It's frustrating when we try to open markets to products and we can't get the investment we need to support the production."

Three businesses surrounding Manaus, owned by distant relatives of each other, provide evidence to support the argument that sustainable development can't be more than a site-specific solution to deforestation. These industries don't provide enough employment, they are infrequently profitable, especially without government incentives, and the lifestyle they afford matches up poorly against the amenities available in urban areas. Braga acknowledges as much, but argues that these small industries have to survive to dissuade rampant migration from rural areas even if they require subsidies. The ancillary benefit of the public investment—the presence of civil society that arises where there is gainful employment—justifies propping up unprofitable businesses. Local industries harmonious with the environment need to be sustained, he says, even with incentives. The lesson of the past is that the Amazon abhors emptiness, as it invites lawlessness. Accepting that as a given, Braga has supported a benign presence in many areas through promotion of indigenous industries.

Asher Benzaken runs two fisheries about a half hour from the Manaus city center. He finds it amusing that his trade is seen as salvation for the forest. "Not this business." He laughs. "Maybe the governor's talking about food farms. This one is much too difficult." Benzaken exports aquarium fish. His two facilities consist of dozens of pools of river water, each holding hundreds of fish, which are counted, catalogued, and fed before they are exported. He employs forty-five people at the two sites, and buys from one thousand hand-net fishermen in Barcelos, about three hundred miles up the Rio Negro. Benzaken's buyers travel through that region buying up collectable fish and then shipping them by boat to Manaus. In his monograph, "Social and Economic Change in Amazonia: The Case of Ornamental Fish Collection in the Rio Negro Basin," Gregory Prang describes how the ornamental fish industry "plays a key role in the local economy," even though the output is relatively small—about two million dollars of exports from the entire state of Amazonas in 1998, the year of his study. He observed that there are "large kin networks" involved in the locating, trapping, selling, and transporting of the fish—primarily the cardinal tetra variety. The industry provides generational continuity and stability to Barcelos, which is a key anchor along the Rio Negro and which is the type of outpost that Braga envisions maintaining.

"As a business, it's okay in terms of profit," Benzaken said, and shrugged his shoulders. "But you're not going to build an economy on it. How many aquariums do you think there are in the world? We also have very tight controls on the fish we can export. There's a list of one hundred fifty fish, no more. The others are protected. Then, we have great difficulty exporting from here. There aren't many flights from Manaus. We have to tranquilize the fish and time it just right, or we will send out a shipment of dead fish."

Benzaken's cousin by marriage, Daniel Amaral, makes a living selling forest products: nuts, seeds, and small vials of oils. He also finds it amusing that his primitive industry is now seen as a model for future development. "It's an interesting business, but these are not soybeans, you know. My business is essential oils, perfume oils. For a time we sold a guarana extract to Coca-Cola, but we lost that business because of a competitor who had political connections. You say the government wants to promote these businesses? Then they shouldn't play favorites and subsidize one factory over another."

Amaral employs forty people at his facility. Boatloads of nuts and seeds are brought up the small stream outside his plant; they are shelled and dried, then the distillation and extraction process takes place. The site resembles a large chemistry experiment, with its gleaming tanks interconnected by a web of pipes. "We invented every process we use here," he said matter-of-factly. "No two nuts are alike. No two processes are alike. So it's difficult to find machines that can perform more than one function. And it's difficult to do anything in great volumes."

Amaral's family has a long tradition of extracting and exporting rosewood, the essence of Chanel No. 5 and other perfumes. The French companies initially bought their rosewood from French Guinea, but they depleted that supply and turned south to Brazil, where rosewood soon acquired the status of mahogany, eventually ending up on Brazil's endangered species list. Now, with government support, small rosewood plantations with rotating extraction schedules provide the supply.

Amaral's business puts E. O. Wilson's *The Diversity of Life* into practice. Amaral views the biodiversity as a treasure chest, the only constraint often being his inability to devise a use for each of the contents. Recently, he discovered a new scent, only to be later flummoxed by the work of translating the find into profits. "There's a certified forest not far from here, and we've developed a venture with them," he explained. "We study the branches and leaves of what they leave behind and try to extract oils from them. We sent some promising samples to Paris to see if they liked them, too. They did! The problem was that we had to go back and find the tree again. We essentially had to identify a new species of tree in order to find our fragrance."

Amaral said he supports the notion of a Zona Franca Verde, and he considers it a triumph that he "can keep a family working with forest products." But he dismissed outright that his business could be a model for anything other than what it already is. "You have to be a realist. The rest of the world just wants to exploit what we have. They don't want us to have patents. They don't want us to develop the final product here. They want us to discover products, tell them how to use it, what's important about it. Then they'll synthesize it and replicate it. And produce it in volume. They did it with quinine. They did it with rubber. Different century. Same thing."

Another industry on the Zona Franca Verde agenda belongs to Davis Benzecry and his family. They own a jute factory, stuffed into a densely populated area in the older part of Manaus. The Japanese brought jute to the Amazon, and it thrives in this tropical soil. It can grow thirteen feet in six months.

Turning jute into bags that carry coffee beans follows a rudimentary circuit, nearly identical to making cloth. Jute stalks arrive in small boats and are cleaned. The fiber is separated and then taken through processes to soften and narrow it into thread, which is then woven into bags. Coffee beans are best transported in these bags, as plastic bags diminish the beans' smell and taste.

"I remember when I started in 1988, people told me there was no future in jute, because plastic was the future," Benzecry explained as he walked through the warehouse, which clanged with weaving machines and spinners. "But the market disagreed and demanded jute over plastic, and we survived."

His survival, however, is precarious. "First, we have to compete with India and Bangladesh. They can produce jute more cheaply, especially when they receive government subsidies. Because of that, our government adds an import duty to their products. Great. But these countries have taken away our natural market in Colombia. Second, I have to compete with my own government. I have a competitor in a small town not far from here who gets government subsidies. You think that's fair to me?"

Like Amaral, Benzecry also dismissed the notion that his business could be a model for the future of the forest. "Not this. Look, we have an office in Coari and a floating dock near Manacapuru which buys jute from people. But how much bigger do you think we can get? Then we have to deal with subsidies from other governments to our competitors. Then look what happened to our Brazil nut business. On July 4, 2003, the E.U. decided to embargo all Brazil nuts, because there supposedly was some toxin in the shell. One day an entire industry disappears because of a decision made thousands of miles away."

He pointed to the mostly female workforce of five hundred working in the dark room. "These people come to work every morning. They work hard every day. Then someone in Paris decides they don't like our nuts or our jute, and we have to stop giving employment. What do you think the people do then? They will cut a tree and sell it if they need money. And what do the people in Paris say to that?"

He shook his head. "The Europeans wonder why people in the Amazon cut down trees or have turned to drug trafficking. It's very simple. The answer is that they have to eat."

Europeans have landed in the Amazon about a half hour's drive from Itacoatiara, where they have purchased 305,000 hectares of forest for the purpose of making a profit from certified forestry. A Swiss company called Precious Woods, a veteran of successful certified forestry in Costa Rica, owns and operates this forest.

Renato Scoop, the chief financial officer of the facility, conceded that what works in Costa Rica may not work in the Amazon—namely, a plantation concept of forestry. "It's very difficult to use a diverse forest as a source of profit. There are economies of scale which make the process very expensive," he explained. A visit to a well-inventoried hectare of certified forest showed why. In just this small area, there were several hundred different species of tree, each with a different cycle for regrowing and each, of course, with unique economic value fluctuating with the market. While it's easy, based on the market and on known expenses, to decide when to cut a hectare of only eucalyptus trees, it's much harder to make that decision on a tree by tree basis for a farm larger than Manhattan.

Scoop relies on computer printouts that identify the location of every single tree in the sixty-seven thousand hectares already catalogued. In addition, he and his teams make decisions about extraction strategies for the trees so they don't destroy other species on their way out. This decision, in turn, is determined by the location of the network of red-clay access roads zigzagging through the property.

The project employs 850 people who, together, have a role in every facet of the journey from a seed to a table. There are those who identify the trees, those who cut them, those who plant the replacement trees, those who haul the fallen trees to the sawmill, those who collect the sawdust and take it to the power plant that runs the machinery, those who saw the wood, those who manage the veneering process, and those who transport the finished wood to the port. There is no other integrated process like it in the forest.

But to think that this system can be replicated to provide the backbone to the Zona Franca Verde is to envision a world of limitless resources. If this is how much capital and care is needed to sustain a single forest of sixty-seven thousand hectares and harvest it in an environmentally friendly manner, this system on a larger scale simply can't be done. It's a lot easier to show up with a chain saw in the middle of the night. "People don't pollute [or deforest] because they like polluting," the economist Ronald Coase has said. "They do it because it's a cheaper way of producing something else."

EPILOGUE

—

As accessible as this exotic world has become, as much as technology has invaded Eden, as pessimistic as we were about deforestation and climate change, as optimistic as we were that the Amazon is better off than it was twenty-five years ago and that in this fairly nasty world there's still hope for good things to happen—nature still humbles. Crossing through the meeting of the black waters of the Rio Negro with the brown waters of the mighty Amazon as the sun sets can make anyone feel small and insignificant, a tourist on the planet.

We could barely make out the darkened wisps of the tops of trees on each side of the river as we listened to the *tut-tut-tut* of the barge moving slowly against the current of waters that had begun their journey a continent away. We could see the silhouettes and deck lights of ocean tankers heading another thousand miles upriver to call on Iquitos in Peru. Dugouts with stacks of bananas weighing down the sterns came alongside the barge just so the shirtless paddlers could shout "*Oi.*" Tourists with video cameras went by in small motorboats on their way to look for alligators dwelling on the river's edge. Groups of exhausted passengers, some having traveled two weeks without a hot meal or a toilet, stood sentry on the prows of river vessels anxious to reach solid ground.

We had our own sense of anticipation as we headed for the south bank to travel the five-hundred-mile length of BR-319 to Porto Velho, the once-paved road that had degenerated from disuse into a rutted jungle path. For some reason the road was ignored by settlers and developers after it had been paved in the 1970s. Rumors persisted that it was about to be repaved. If true, and if history is an accurate prognosticator, then millions of hectares of intact rain forest of the western Amazon are going to be put into play. It is unfathomable to believe this road will be ignored again.

Small trucks took up most of the space on our barge. The lights of Manaus faded to a dim aura, fittingly. Decades of progress have transformed the north bank and overpassed the south, where the arrival of a barge qualified as a spectator event. Throngs of people had turned out among the food stands, occasional plastic card tables and chairs, and blaring music. Some of the trucks had trouble handling the steep hill leading up from the river, and there was no shortage of boys and men shouting instructions—"forward, reverse, forward"—and helping push.

The phenomenon of road building in the Amazon has deflected either resources or attention, or both, from an obvious failing of infrastructure: adequate docking facilities. Although building reliable structures is problematic where rivers rise and fall so dramatically from season to season, there's little evidence of effort to modernize river commerce among the towns. For a fraction of the reported 139 million reais the federal government has committed for the years 2004 to 2007 for repaving BR-319, port infrastructure could be improved to facilitate trade. For an inexplicable reason, rivers are not treated as viable highways, except by Blairo Maggi.

The road to Careiro Castanho turned out to be passable, although a few hundred yards of mud and rocks interrupted unblemished pavement at regular intervals. Reportedly, thirty years ago owners of a barge company showed up with heavy equipment shortly after the road was paved, and tried to destroy as much of it as possible. The road had broken their monopoly on transport from Manaus to Careiro Castanho, and they wanted the monopoly back. Enough shattered rock remains today to make it a challenging journey.

In Careiro Castanho, a crowd of teenagers loitered in fours and sixes underneath the few streetlights. They leaned on bicycles, strolled and held hands, sat on benches and talked. The two open-air cafés, lit by strings of Christmas tree lights, were full, serving beer, sodas, and grilled fish. Half of the tables were taken by mating teens trying to gossip above the wailing music, and the other half were taken up by sports fans watching yet another replay of the Olympic women's soccer game between the United States and Brazil. Many Americans probably envision folks in a small Amazonian town walking around nearly naked balancing bunches of fruit on their heads. Yet even in a remote place like Careiro Castanho, a parallel universe exists. What happens here on a Thursday night happens in most rural communities across America.

Beyond Careiro Castanho lies a land that few people know. The Manaus newspaper carried a story about someone who got stuck in the mud along the road for four months. From time to time, trucks fall through bridges, where they decompose over decades. Panthers patrol long segments of the road. When people in Careiro heard we were going on the road, they wanted to know why, and, without exception, they wished us good luck. Most of them had never been a mile south of the town. No one said, "I wish I were going with you."

To go south of Careiro late that night, we waited on the bank of a small river and honked the horn over and over until a distant voice shouted *"Oi"* from the darkness a hundred yards away on the other side. The *"oi"* was followed by a price list per passenger, per car, per motorcycle, and the negotiations took place in the quiet darkness, back and forth, until we struck a deal with the man who owns the only barge in town.

Then, silence. Nothing happened. We stood there wondering if the voice had gone back to bed, if the deal we'd struck wouldn't be effective until first light. There might be a hotel in Careiro; then again, we hadn't seen any. There was no barge back to Manaus that late. We remembered something we'd heard a long time ago: "There are no schedules in the Amazon." Perhaps the voice had to finish dinner. What did it matter if ten, twenty minutes or two hours passed? The journey itself was the destination.

The barge finally came to life after almost an hour, and we continued on the road toward a small village called Tupana twenty-five miles away. There were no lights, no houses, no sense of civilization, and strangely we found the finest pavement in all of Brazil. It was like coming upon Machu Picchu, an engineering marvel in the jungle. Twenty-five miles without a pothole, without a blemish. Why? To connect nothing to nowhere? How many educations and meals could this thoroughfare have paid for? How many politicians got rich off this pavement?

Then Tupana, and it was very much night, and another small river to cross and no one about. So, again we searched for a voice. *Honk. Honk.* Eventually we heard an "*oi,*" we negotiated a price, and we waited. The Tupana barge is a Rube Goldberg device, resembling Noah's ark pushing a metal platform. As we crossed the 160 feet or so, we saw a bridge that looked like a casualty of the Kosovo war. The barge man said a truck had fallen through it. He was smiling, so we wondered how unhappy he'd been when the competition had collapsed.

Then the road turned bad, so bad it didn't really exist. There was a ribbon of red clay in the headlights, curving up a hill to the right and—*whoompf*—we fell into our first crater of mud. The wheels squealed, the car heaved back and forth over and over, and finally we emerged. There was some pavement here and there but eaten away from the sides to the middle, and we were desperate to stay on whatever narrow strip of asphalt remained. There were no houses, no lights, nothing outside except early-morning darkness. Driving required the skills of an ER surgeon to avoid the mud, find the little remaining pavement, find the dry path in the jungle where the road might have been. Then it was a good time to rest.

Tens of thousands of people left their homes in the northeast and south of Brazil in the late 1970s and early 1980s and traveled roads like this in search of a new life. They slept in places like this, moved forward along roads like this, enduring snakebites, malaria, hunger, heat, and the grime of a journey. They had to feed and protect the children they had brought along. They had to try to find a home somewhere in this wilderness, because they had no home to go back to.

Making the decision to leave a known life, preparing for an expedition into the unknown, and surviving the rigors of the rain forest—these experiences have shaped the character of the people who have come to occupy the Amazon. Optimism and self-reliance have been planted here, although the people's clashes with natural elements and man-made circumstances have diminished these traits' influence. Brazil, as it contemplates productive uses of these human and natural resources, faces the challenge of how to rejuvenate and foster the spirit that motivated this migration, and how to direct it in favor of a sense of common destiny. A tall task for any country, although an opportunity few countries have anymore.

Remembering the earlier migrations brought to mind the startling absence of signs of any occupation—present or past. This BR-319 was paved when BR-364 was a thousand-mile mud hole; no one came here, yet thousands risked the mud on that road farther south. If there had been settlers, they were all gone now without a trace. Pioneers vote with their feet. Why was no one living along BR-319? Was the soil poor? Was the climate not favorable? Something must have been wrong with BR-319. Otherwise, this land would have been settled years ago.

Money has now been set aside to repave this road. Responding to the business community of Manaus, where he once was the mayor, the Minister of Transportation Alfredo Nascimento is pushing the plan. The commercial interests in Manaus have long coveted a land link with the rest of the country, even if it is interrupted by a slow barge ride across the river. But didn't they already have that link? Why spend all this money now to repave it? Why not spend the money repairing BR-364, which is badly in need of maintenance and already is a rich vein of commerce? Why not use the money for better planning and support for the soy highway BR-163, soon to be paved? Or educating Socorro's kids in Altamira? Or feeding the kids along the Xingu River? Why spend so much to help the unseen? Is this the wisest allocation of resources? In this young democracy, how should this decision be made? When this road is repaved, it will put a knife wound into the vast untouched forest of the western Amazon. The land Carlos Marx said that "no one can get" to will be coveted, claimed, and unalterably changed.

At first light, after a few miserable hours of sleep in a storage trailer by the side of the road, we continued. At the small settlement of Igapo-Acu, a handful of small houses, we came upon another tributary. We discovered a new law of the jungle: the barge is always on the other side. A young mother was washing her children's clothes in the water. Young kids with a makeshift fishing pole were shouting into the water, urging the fish to come visit them. The barge slowly made its way across the water.

"This is where the pink dolphins live," a boy named Elielzo on the barge said. "They will come to see you when I call them. If you want."

As they really do exist, they're not quite the Loch Ness monster, but the pink dolphins have mythical status in the Amazon. To some, the dolphin becomes a young man who seduces unmarried girls; to others, it becomes a lovely woman who leads young men to the water and disappears with them. In some places, women who go swimming while menstruating are said to be made pregnant by the pink dolphin, and the offspring have magical powers. Unfortunately, the dolphins' number is shrinking; their parts are prized for their supposedly aphrodisiacal and magical powers.

"Here they will come to you and you can touch them," Elielzo said.

And so when the barge reached the other side, he went onto the pilot boat and leaned over into the water. He splashed. Nothing. He splashed again. Nothing. He pulled a small fish, the size of an index finger, from his pocket. "*Boto*," he called softly. "*Boto*." *Nada*.

"They're really here," he said. "Truth." He began to flap the surface of the water with the bottom of his flip-flop. Some of the children who were fishing did the same. "*Boto. Boto*," they called out.

And then they came. Two of them. Pink like a rose. It's said they're pink because their capillaries are close to their skin. They're tranquil, too. Supposedly their brain capacity is 40 percent larger than humans'. This is either proven or disproven by the fact that they trust us to pet and feed them. If the road is repaved, they will die soon afterward.

The jarring returned soon enough, as did the washed-out bridges. Occasionally, we had to get out of the car to lighten the load across the bridges. There were no signs of humanity. The jungle had closed in over the remnants of the road that were like islands of asphalt in seas of red clay mud. There was no escaping the mud holes. The wheels squealed as the car shifted back and forth, churning its way out of the slime, and we hoped this wouldn't be the crater that swallowed the car for hours, days, or months.

Paulo Nazare dos Santos got stuck for ten months. He lives at Kilometer 350 on the road. With nothing to do for ten months, he built a house. "It's all Amazonino's fault," he confided. "Yes, it is."

How did Amazonino Mendes, the former governor of Amazonas, get Nazare stuck in the middle of nowhere? "I lived in Humaitá. This was about seven years ago. Amazonino came there and said anyone who owns land on BR-319 should go back. The road will be paved. I didn't own land, but the man I worked for did. But he had malaria. He told me to take his land."

Nazare's timing was bad. Seventeen bridges washed out in the rainy month after he arrived. "I lived on fish and farinha. It was very lonely." Now he has a wife and three children. He's forty-one years old, and he raises turtles and *tucunare* fish in a holding pond. He gets paid for helping to maintain the Embratel communications towers and the bridges in the vicinity of his house.

Then we have more of our ceaselessly difficult journey. When it rains, no one can get through. Not only does the mud become totally unnavigable, but gauging the depth of any puddle is impossible. Unexpectedly, a stretch of about half a mile of smooth asphalt provides relief. The jungle has been cleared for several hundred yards on each side of the road. Supposedly, some farmers from Mato Grosso built this as a runway for their plane. We hear they have made plans to rip away the jungle as far as the eye can see. Soy is coming.

Someone is walking along the side of the road, a Davy Crockett–like rifle wrapped over his shoulder. He says his bicycle got damaged when he hit a panther. He shot the panther. It would be another three days of walking before he arrived home.

The next goal becomes a house at Kilometer 480, where "The German" lives.

As we get closer, the bridges get better, sturdier, and they are marked by white wooden arrows, which prove to be great travel aids on a dark road surrounded by dense jungle. The German did his job well.

Gilson Schreider lives with his wife and five very blond kids. The main room of the house has reading aids on the wall; they homeschool the children. "When we first got here, we sent our cows to Porto Velho, and it started to rain," he explained, justifying his support for the repaving. "They all died in the trucks, stuck on the road. Now we don't send any once September arrives."

He's a traveler's friend, offering *cafezinhos* to the weary, television to the uninformed (he's got a generator), and anecdotes of those who came before. "Be very careful from here," he warned, as if this was news. "The road is very bad. You see that car there." He stood in his doorway, pointing to one of several in his yard. "Someone got stuck in the mud and walked away. When the rains stopped I got the car. He never came back." He said he gets spare parts from trucks that tumble from the road and die in the mud.

There was a moment's pause at Kilometer 500, where an abandoned gas station provided the last roof until Humaitá. Stay the night here, or carry on? At best, Humaitá was another seven hours away, meaning an arrival past two A.M.

Humaitá sits on the TransAmazon Highway. Under most circumstances, no one would go there. Its existence is happenstance, a settlement that grew up around a road that no one really wanted in the first place. But now it loomed like Altamira in *Bye Bye Brazil*, a toilet of a town disguised as a destination paradise.

The road would not bend to the willing. The headlights were of marginal utility. All the splashing in mud covered them with an opaque film of thin clay. We simply hit walls of blackness. So, we stopped and walked about, searching for the road. Walked several hundred yards to make sure it was the road and not a path to nowhere. When lucky, we didn't step into knee-deep puddles of water. Sometimes we ended up down an embankment, toward a narrow river, and realized we must have missed the bridge. Driving in reverse gear up an incline—pushing helped—was yet another challenge on this exhausting ride. But it was a diversion, this search for the road, returning to the car, and driving until it disappeared again. And it kept us awake; that and the violent, never-ending stream of whiplash.

We came upon the intersection of BR-319 and the TransAmazon Highway when the early morning felt like a heavy damp blanket. The sign pointed to the right, where the town of Labrea lay, another day's journey with any luck. To the left and only eighteen miles and an hour away Humaitá awaited, an oasis of twinkling lights, a hotel among them. On this particular night Foster Brown was wrong: Humaitá would be the center of the universe.

ACKNOWLEDGMENTS

—

Researching and writing this book has been a labor of love, not only because of the great subject we've had to address but, more important, because of the great support we've received in this endeavor. The order in which names appear in this section is random, not by importance. Everyone has been important to this project.

We were fortunate to find Amy Rosenthal one summer day in 2003, helping Foster Brown (who helped us a ton) and his team in their work along the Road to the Pacific. We embraced Amy as soon as she returned to the United States, pinned the "researcher" badge on her, and reaped the fruit of her extraordinary work. Amy organized three significant trips to the Amazon and brought back hundreds of pages of interviews, impressions, insights, and photographs. Vera Reis, Amy's pal from Rio Branco, was at her side throughout these trips, and we are the beneficiaries of her fine work product. If there's a hall of fame for researchers, Amy and Vera belong there. Phil Tucker went with Amy and Vera on one trip, and he, too, gave us valuable material to work with.

The Brazilian embassy in Washington, D.C., told us early on that we could count on their support. They meant it. They delivered repeatedly. They provided access that we simply couldn't have gotten on our own.

In the process, we again became great admirers of the excellence of the Brazilian foreign service and great friends with many of its representatives. Ambassadors Rubens Barbosa and Roberto Abdenur were always looking out for us and eager to help, and we are grateful to them. Marcos Galvão, who served as their deputy, became our close friend and a key supporter. We also are grateful to Cafredo Texeira, Roberto Goindach, and Carlos Villanova. The de facto diplomatic corps in Washington also showed up at our side whenever needed, especially our neighbor Paulo Sotero, as well as Luis Bittencourt and Thomaz Costa.

In Washington, we also found great support at the World Bank from Nils Tcheyan, John Redwood, Harold Rosen, Bob Schneider, and Adriana Moreira (whom we first found in Brasília). Those in Washington who have shown care and concern for the Amazon over the years and who never hesitated to help us include: Riordan Roett, Steve Schwartzman, Russ Mittemeir, Tom Lovejoy (who has been extraordinarily helpful since the first interview for the first book, almost thirty years ago), Gustavo Fonseca, and Larry Small. Also, we pledged never to write a book without guidance from the "man who knows everyone," and H. P. Goldfield came through for us once again. We never could have done any of this without his colleague Joel Velasco, who adopted this project as his own and was always there for encouragement and support.

There's a crew of folks who researched for us, thought with us, encouraged us, and read and edited for us, and we are grateful to them all. They include Beatriz Portugal, Dr. Mercia Flannery, Brian Sexton, David Maraniss, Mary and Larry Ott, Arne Sorenson, Bradley Clements, Alex Walker, Keri Fiore, Kevin Mead, and Ann McDaniel. Chick Hill led one of the greatest expeditions the Amazon has ever seen, and we thank him for his support and friendship, and we salute the men in the cockpit—Rich Boyle and Steven Brown.

Many of the people in Brazil who helped us will find their names already in the book, and we want them to know that every time they see their names, they should be hearing a big *obrigado*. Additional *obrigados* go to the entire Benchimol family, especially Jaime, Anne, and our traveling pal, Denis Minev. President Fernando Henrique Cardoso, Everton Vargas, Wanja Nobrega, Ana Cabral, Andre Lago, Ruy Amaral, Leticia Meirelles, Olga Lustosa, Cloves Vettorato, Virgilio Viana, Ane Alencar, Toby McGrath, Dan Nepstad, Juliana Moreira Lima, Olimpio Cruz, Etel Carmona, Blairo Maggi, Jorge Viana, and Eduardo Braga— we hope these thanks find them happy and healthy and continuing to do the good things they do. João Carlos Meirelles reminded us that it's possible to grow smarter as we grow older. We constantly worry about our colleague Lucio Flavio Pinto, and we wish him serenity and security; he already has wisdom. Ambassador Donna Hrinak in the U.S. embassy in Brasília and Eric Stoner, a fixture in that community, gave us their support.

In our offices, Chris Mead showed that he's not only America's finest trial lawyer, but he's also America's finest law partner and a true friend. The team of Hanoi Veras, Lisa Manning, and Mary Jane Snyder provided great support and cheerleading day in and day out. Brian Duffy was supportive from the very first day and bailed Brian out with his usual grace whenever it was needed, and Mort Zuckerman's encouragement has meant a lot.

On the writing front, we start with the memory of Sara Stein, who guided us through our first book and inspired in us an enduring interest in the Amazon and the confidence with which we could chronicle it. We're always grateful to our lifelong friend Ann Godoff, who, for the second time, made sure we were all right. David Ebershoff started working with us, and we appreciate his support and wish him great success in his writing career. Ben Loehnen stepped into David's role, and we still can't believe how lucky we were that he wasn't busy that day. Ben put it all together; he's the very best! Many thanks to Bara MacNeill, our copyeditor. Our agent and friend forever, Rafe Sagalyn, was always there, as usual.

And in the end, we hug our families who hung in there for us: our eternally supportive parents, the inimitable Dania, the indomitable Patti, and the kids to whom this book is dedicated.

Obrigado and thank you to all.

NOTES

—

Interviews and Travel

The interviews in the book were conducted on multiple visits to Brazil and are referred to as "Authors' Interview." Where noted, a few interviews are taken from our first book, which was based on two visits, each of about four months, at the end of 1980 and again at the end of 1981. The other interviews were conducted during our visits from 2003 to 2005. Mark made eight trips to Brazil during this time, Brian made six, and our researchers, Amy Rosenthal and Vera Reis, made three. We have adopted all of these trips as our own for the narrative use of "we." An interview by any one of us is referred to as "Authors' Interview." In several instances, where noted, the same person was interviewed more than once during this two-year period.

Non-Brazilian Sources

Our Brazilian friends correctly pointed out that most of our source materials were written by North Americans for a U.S. audience. By concentrating on these sources, we do not intend to make any statement about the comparative worth of these sources versus the sources that are available from Brazilian authors. The choices we made were primarily dictated by our preference to research in English.

Definition of the Amazon

We apologize for the loose and interchangeable use of the terms "the Amazon," and "Amazonia." The precise definitions of these terms are subject to debate beyond wordsmithery. For example, Brazil's "Amazon" refers to an administrative region comprising nine states—Acre, Amapá, Amazonas, Pará, Rondônia, Roraima, Mato Grosso, Maranhão, and Tocantins. This is an artificial definition, as the dense rain forest in the west of Amazonas is distinct from the *cerrado* of the south in Mato Grosso by most standards—climate, agronomy, biodiversity, and so on. Business people and politicians in Mato Grosso advocate removing at least part of their state from the definition of "Amazon," which in turn would reduce outside scrutiny of their activities, as they would no longer be included in annual deforestation figures. In June 2005, responding to a request from the secretary-general of the Amazon Cooperation Treaty Organization for assistance in defining the geographical boundaries of Amazonia, a group of scientists produced "A Proposal for Defining the Geographical Boundaries of Amazonia" (available at www.ies.jrc.cec.eu.int). That report notes that in defining "Amazonia," one needs to take into account hydrographical criteria (defining the area as that where watersheds drain into the Amazon River), ecological criteria (defining subregions based on ecological similarities), or biogeographical criteria (defining the region as the "known historical extent of the Amazon lowland rain forest biome in northern South America"). We feel that these important distinctions would be distracting in the context of our reporting, but we alert our readers to the issue.

CHAPTER ONE

The several references to *1491* by Charles C. Mann in this chapter should reflect the importance we attach to his work and the admiration we have for it. Mann's article in *The Atlantic Monthly*, April 2002 (also titled "1491"), provided the initial intellectual inspiration for our decision to have another look at the Amazon. (We also relied on other Mann articles that predated his book: "Earthmovers of the Amazon" and "The Good Earth: Did People Improve the Amazon Basin," both in *Science*, 2000.) We credit Mann for choosing Betty Meggers and Anna Roosevelt as the exemplars of the two sides of the debate over the meaning of the history of the Amazon. We also appreciate Mann's occasional e-mails of clarification and support. We interviewed Meggers for our first book and conducted a phone interview with Roosevelt in March 2004 for this book. Roosevelt insisted that a visit to Monte Alegre with anyone other than Nelsi Sadeck would be a waste of time. We never tested that theory; we put ourselves in the capable hands of this enthusiastic and resourceful guide during our visit to Painted Rock Cave in May 2004. A short article in the March 2003 issue of *GSA Today* by William R. Brice and Silvia F. De M. Figuieroa provided helpful information about an early visitor to Monte Alegre: *Rock Stars: Charles Frederick Hartt— A Pioneer of Brazilian Geology.* Our quote from Robert Goodland comes from an interview for our first book and appears in that text. We found Susanna Hecht's Horace Albright Lecture at the University of California in 1993 to be an excellent summary of the current state of knowledge about the peopling of the Amazon. Hecht's lecture, as well as her 1989 book with Alexander Cockburn, *The Fate of the Forest,* make us wish that she continue to write about the Amazon; her work has been varied in subject matter and extraordinary in quality. The reference to Andrew Revkin's and Jonathan Kandell's mentions of the *terra preta* findings in their books should be read as praise for the thoroughness of their research.

5 *"This is a magical place"*: Authors' Interview, May 2004.

5 *1960 article:* Helen Constance Palmatary, "The Archaeology of the Lower Tapajós Valley, Brazil," Transactions of the American Philosophical Society N.S., 50: 3 (Philadelphia, 1960).

5 *"at the time"*: A. C. Roosevelt, "New Information from Old Collections: The Interface of Science and Systemic Collections," *CRM* No. 5, 2000: 25–30, p. 26.

6 *The age of the pottery:* Ibid., p. 25.

6 *"pile of yellow pages"*: Ibid., p. 26.

6 *"Wide and shallow"*: Charles C. Mann, *1491: New Revelations of the Americas Before Columbus* (New York: Alfred A. Knopf, 2005), pp. 293–94.

7 *"30,000 stone artifacts"*: Roosevelt "New Information," p. 28.

7 *scientists had adopted:* A. C. Roosevelt et al., "Paleoindian Cave Dwellers in the Amazon: The Peopling of the Americas," *Science* 272: 373–84 (1996).

7 *"Even modern efforts"*: B. J. Meggers, "Environmental Limitations on the Development of Culture," *American Anthropologist* 56: 801–24 (1954).

8 *the soil in the Amazon:* P. Richards, *The Tropical Rain Forest: An Ecological Study* (New York: Cambridge University Press, 1952).

8 *"with all its wonderful intricacy"*: B. J. Meggers, *Amazonia: Man and Culture in a Counterfeit Paradise* (Chicago: Aldine-Atherton, Inc., 1971), pp. 157–58.

9   *"The purpose of this study"*: R.J.A. Goodland and H. S. Irwin, *Amazon Jungle: Green Hell to Red Desert?* (New York: Elsevier Scientific Pub., 1975), p. 1.

9   *settled groups*: D. W. Lathrap, *The Upper Amazon* (New York: Praeger, 1970).

10   *the environment inhibited*: Robert L. Carneiro, "A Theory of the Origin of the State," *Science* 169: 733–38 (1970); and "Slash and Burn Agriculture: A Closer Look at Its Implications for Settlement Patterns" in F. C. Wallace (ed.), *Man and Culture* (Philadelphia: University of Pennsylvania Press, 1960), pp. 229–34.

10   *"extensively modified"*: C. D. Sauer, "Man in the Ecology of Tropical America," *Proceedings of the Ninth Pacific Science Congress* 20: 104–10 (1957).

10   *vast areas of the modern Amazon*: William M. Denevan, "The Pristine Myth: The Landscape of the Americas in 1492," AAAG 82: 369–85 (1992).

10   *beginning to publish the results*: A. C. Roosevelt, *Moundbuilders of the Amazon: Geophysical Archaeology on Marajó Island, Brazil* (San Diego, Calif.: Academic Press, 1991).

11   *Roosevelt, in turn*: Mann, *1491*, p. 297.

11   *"upon the majesty"*: Susanna Hecht, "Of Fates, Forests and Futures: Myths, Epistemes, and Policy in Tropical Conservation," Horace Albright Lecture, University of California at Berkeley (1993); available at www.cnr.berkeley.edu/forestry/lectures/albright/1993hecht.html.

11   *"You can think of it"*: Authors' Interview, October 1980.

12   *"The new picture"*: Mann, *1491*, p. 311.

12   *most sophisticated society*: Michael Heckenberger, "Remains of Cities Found in Amazon Basin," Associated Press, September 23, 2003.

12   *"Adherence to"*: B. J. Meggers, "The Continuing Quest for El Dorado: Round Two," *Latin American Antiquity* 12: 304–25 (2001).

13   *pure, primeval, or primitive*: phraseology from Hecht, "Of Fates, Forests and Futures."

CHAPTER TWO

We wanted to minimize the discussion of the "discovery" of the Amazon, because it's been written about so many times. Mann and Hecht have led the way in suggesting that Carvajal deserves another look, which provides an opportunity to try to separate fact from fiction in an account that, for four hundred years, had been dismissed as fiction. Shoumatoff's piece in *The New Yorker* on Bates is a great character study of an underappreciated scientific figure. For us, Shoumatoff never will be unappreciated. When we read his *The Rivers Amazon* in 1978, we realized that the place wasn't that foreboding and was on the verge of newsworthiness. That book and his subsequent writings have made the Amazon accessible for us. We quote several times from *The World Is Burning*, and we also relied on his comprehensive description of the building of Brasília in *The Capital of Hope*. We've viewed all of the fact summaries of the physical description of the region with suspicion and probably found the most comfort in Revkin's background information in *The Burning Season*. In 2006, the word "biodiversity" regularly appears in our vocabulary, but we haven't lost sight of the extraordinary work of E. O. Wilson to make it so. Peter Raven, who coined the term with him, was very supportive of our efforts twenty-five years ago, and sponsored our visit to the Missouri Botanical Garden. We've never met Donald Perry or Terry Erwin, those who live at the top of the trees; they remind us that there still are many frontiers to learn about in plain sight.

15 *History has bestowed:* reliance on *1491*, Alain Gheerbrant, *The Amazon: Past, Present and Future* (New York: Harry N. Abrams, 1992), pp. 14–27.

16 *"The Amazons":* Ibid., p. 27.

16 *"We saw the":* Ibid., p. 13.

17 *"squadrons on the riverbank":* Susanna Hecht, "Of Fates, Forests and Futures: Myths, Epistemes, and Policy in Tropical Conservation," Horace Albright Lecture, University of California at Berkeley (1993); available at www.cnr.berkeley.edu/forestry/lectures/albright/1993hecht.html.

17 *another written account:* Cristóbal de Acuña, *The New Discovery of the Great River of the Amazons, 1639* (Madrid: Royal Press, 1641).

18 *La Condamine wrote:* See Charles-Marie de La Condamine, *Abridged Narrative of Travels Through the Interior of South America* (Barcelona: Editoria Alta Fulla, 1986); *Descending the River of the Amazons* (Paris: François Maspero, 1981).

18 *In the mid-nineteenth:* Alex Shoumatoff, "A Critic at Large, Henry Walter Bates," *The New Yorker*, August 22, 1988.

19 *Wallace founded:* A. R. Wallace, *A Narrative of Travels on the Amazon and Rio Negro, with an Account of the Native Tribes* (London: Ward, Lock & Co., 1853).

19 *When Bates read:* Shoumatoff, "A Critic at Large."

19 *Bates's highly readable:* Ibid.

19 *Spruce's contribution:* R. Spruce, *Notes of a Botanist on the Amazon and Andes*, 2 vols. (London: Macmillan, 1908).

19 *Wallace, an uncommonly generous:* Wallace, *A Narrative of Travels*.

19 *"South America":* Theodore Roosevelt, *Through the Brazilian Wilderness* (New York: Charles Scribner's Sons, 1914).

20 *many titles attest:* Robert Churchward, *Wilderness of Fools: An Account of the Adventures in Search of Lieut.-Colonel P. H. Fawcett* (Oxford: Routledge, 1936); William LaVarre, *Gold, Diamonds and Orchids* (New York: Fleming H. Revell Company, 1935); Jørgen Bitsch, *Across the River of Death* (London: Souvenir Press, 1958); Sasha Siemel, *Jungle Wife* (New York: Doubleday, 1949); Julian Duguid, *Green Hell: Adventures in the Mysterious Jungles of Eastern Bolivia* (New York: The Century Co., 1931); James Foster, *Lost in the Wilds of Brazil* (New York: The Saalfield Publishing Company, 1933); F. W. Up de Graff, *Head Hunters of the Amazon* (New York: Garden City Publishing Co., Inc., 1923).

21 *Amazon basin spreads:* The difficulty in grasping numbers can be seen by comparing the reliable Andrew Revkin, *The Burning Season: The Murder of Chico Mendes and the Fight for the Amazon Rain Forest* (Boston: Houghton Mifflin Company, 1990), p. 7 ("the Amazon River basin . . . is a shallow bowl covering 3.6 million square miles"), with the equally reliable Public Broadcasting System ("within the 2.5 million square miles of the Amazon basin . . ."), found at www.pbs.org/journeyintoamazonia.

21 *"6,762 kilometers [4,203 miles] in length":* Commission on Development and Environment for Amazonia, *Amazonia Without Myths* (New York–Hong Kong: Books for Business, 2001), p. 11.

21 *Hardly a reference:* The physical history is drawn from many sources, but special recognition is given to Revkin, *The Burning Season*.

22 *The river draws:* Ibid.

24 *"163 species":* Thomas W. Kral, "Biodiversity Discovered," available at www.sovereignty.net/p/land/kral-insect.htm.

24 *In the late 1980s:* Interview with Raven, available at www.pbs.org/saf/1106/features/raven.htm.

CHAPTER THREE

We mention the Cardoso administration in this chapter, although do we not include any interviews with President Fernando Henrique Cardoso. In fact, we had the pleasure of having a lunch and dinner with him at the Brazilian embassy in Washington in 2005 and appreciate very much his encouragement. Cardoso was the medicine that Brazil's young democracy needed, providing economic stability and, at the top, probity and intelligence. Paulo Sotero, the Washington, D.C., correspondent for *O Estado de S. Paulo,* has made the excellent observation that President Cardoso in retirement has also served the country well in the role of elder statesman and as a calming influence. We were fortunate to have *The Brazilians* as a resource, suggested by our Brazilian friends as an insightful character study by an outsider. In this chapter, we drew upon our several trips to Brasília, a city that has blossomed over time into a very livable environment.

25 *"The problem":* Authors' Interview, May 2005.

26 *"the country of the future":* Financial Markets International, Inc., "Developing Markets Today"; available at www.fmi-inc.net/fmi_feb_1999.pdf.

26 *It acquired:* Joseph A. Page, *The Brazilians* (New York: Da Capo Press, 1995).

26 *Brazil's economy:* The Evian Group Webletter, March 2004; available at www.cvm.gov.br/port/public/publ/490.pdf.

27 *"The Flat World":* Thomas L. Friedman, *The World Is Flat: A Brief History of the Twenty-first Century* (New York: Farrar, Straus and Giroux, 2006).

28 *Portugal was:* Thomas E. Skidmore, "Brazil's Persistent Income Inequality: Lessons From History," *Latin American Politics and Society,* Summer 2004.

28 *one scholar characterized as:* Ibid.

28 *To establish rule:* Page, *The Brazilians,* p. 37.

28 *Even today:* "In Brazil, the Poor Stake Their Claim on Huge Farms," *The Wall Street Journal,* July 10, 2003.

28 *The wealthiest 1 percent:* Estarislao Gacitúa Marió and Michael Woolcock, *Social Exclusion and Mobility in Brazil* (Brasília: IPEA/World Bank).

30 *In his insightful:* Page, *The Brazilians.*

31 *Shortly thereafter:* Rollie E. Poppino, *Brazil: The Land and People* (New York: Oxford University Press, 1973).

31 *"cavalier disregard":* Ibid., p. 77.

31 *"these intrepid":* Ibid., p. 45.

32 *appeared in schoolbook maps:* Bill Donahue, "The Believers," *The Washington Post Magazine,* September 18, 2005.

32　*"the curved"*: Ibid.

32　*Houses were to be:* Ibid.

33　*the Russian astronaut:* John Dos Passos, *Brazil on the Move* (London: Sidgwick and Jackson Ltd., 1973), p. 128.

33　*Instead, Brasília:* Authors' Interview, Father David Regan, Nov. 1980.

33　*"occult forces"*: Page, *The Brazilians,* p. 211.

CHAPTER FOUR

We first mention Charles Wagley in this chapter, a towering figure in Amazon studies. Wagley's *Amazon Town,* written in 1953, provided scholarly and reliable information about everyday life in Amazonia. More important, Wagley proved to be an extraordinary influence on hundreds of students who went to the University of Florida to study with him. We point out the terrific work of Marianne Schmink and Charles H. Wood in *Contested Frontiers in Amazonia,* just one of their many contributions. Their collection of essays in *Frontier Expansion in Amazonia* was very helpful. Dennis Mahar, whose name does not appear in these pages, is another University of Florida product. Mahar's work in the late 1970s at the World Bank, along with Robert Skillings's, served as the road map for our writings about the road known as BR-364. Their young research assistant Nils Tcheyan fortuitously reappeared in our lives years later. President Sarney's dominance of this chapter most likely will stir controversy because he claims credit for Brazil's move toward environmentalism. We make no judgments as to how much or little credit he deserves, although we note that the law of unintended consequences provides a discernible pattern in many areas of Amazon study. We mention Adrian Cowell in this chapter, who has produced a body of fine work in print and film. We appreciate the time he spent with us surveying his experiences.

35　*In 1964:* Shelton H. Davis, *Victims of the Miracle: Development and the Indians of Brazil* (New York: Cambridge University Press, 1977). Davis's book is a fine account of the state of the indigenous people, pointing out that efforts to deal with the issue in Brazil fit into the same "too little, too late" category as efforts in the United States.

35　*Even today:* Scott Wallace, "Into the Amazon," *National Geographic,* Vol. 204, No. 2, Aug. 2003.

37　*"Brazil seems"*: See Charles Wagley, *Amazon Town: A Study of Man in the Tropics* (New York: Oxford University Press, 1976), Preface to 1976 edition.

37   *"The government planned"*: Marianne Schmink and Charles H. Wood, *Contested Frontiers in Amazonia* (New York: Columbia University Press, 1992), p. 349.

37   *"Contrary to"*: Luis Bitencourt, "The Importance of the Amazon Basin in Brazil's Evolving Security Agenda," in Joseph S. Tulchin and Heather A. Golding (eds), *Environment and Security in the Amazon Basin* (Washington: Woodrow Wilson Center Reports on the Americas #4), p. 71. Paulo Sotero, the U.S. correspondent for *O Estado de S. Paulo*, pointed out that the Gore statement is similar to the sentiment expressed by Senator Robert Kasten of Wisconsin at a memorial service for Chico Mendes, when he suggested that the Amazon belonged to the world and not only to Brazil.

37   *"relative sovereignty"*: Ibid.

37   *"The international"*: Ibid.

38   *"Brazil should"*: Alexander López, "Environmental Change, Security, and Social Conflicts in the Brazilian Amazon," *Environmental Change & Security Project Report*, Issue 5 (Summer 1999), p. 26; available at http://www.wilsoncenter.org/topics/pubs/ACF26A.pdf.

38   *We visited*: Authors' Interview, Nov. 2003.

39   *More than eight*: Adrian Cowell's film is *The Decade of Destruction* (United Kingdom, 1990).

40   *In the Amazonian*: Alex Shoumatoff, *The World Is Burning: Murder in the Rain Forest* (Boston: Little, Brown and Company, 1990).

40   *absorbed as much oxygen*: See, for example, Henry Chu, "Deforestation, Burning Turn Amazon Rain Forest into Major Pollution Source," *Los Angeles Times*, June 20, 2005: " 'It's not the lungs of the world,' said Daniel Nepstad, an American ecologist who has studied the Amazon for 20 years. 'It's probably burning up more oxygen now than it's producing.' . . . 'Concern about the environmental aspects of deforestation now is more over climate than [carbon emissions] or whether the Amazon is the "lungs of the world," ' said Paulo Barreto, a researcher with the Amazon Institute of People and the Environment. 'For sure, the Amazon is not the lungs of the world,' he added. 'It never was.' " See also, *Amazonia Without Myths*, p. 6: "When it is alleged that the Amazon produces a high percentage of the oxygen of the planet, the size and importance of the oceans in this regard is

overlooked, and the importance of one tropical region over all other tropical regions is exaggerated. Also overlooked is the fact that a mature forest maintains an almost perfect balance between oxygen production and the fixation of CO_2."

40 *The 1957:* Artur César Ferreira Reis, *A Amazônia e a Cobiça Internacional* (Rio de Janeiro: Edinova Limitada, 1960).

40 *Periodically, this brand:* Shoumatoff, *The World*, p. 50.

40 *In April 2003:* Authors' Visit.

41 *eight Amazon countries:* Larry Rohter, "Deep in Brazil, a Flight of Paranoid Fancy," *The New York Times*, June 23, 2002.

41 *"The irony":* Authors' Interview, April 2003.

41 *"bigger than Belgium":* Shoumatoff, *The World*, p. 340.

41 *"bigger than Rhode Island":* Ibid.

44 *"Chico had":* Ibid., p. 141.

45 *"give us your pollution":* Recounted to us by Paulo Sotero, U.S. correspondent for *O Estado de S. Paulo*.

CHAPTER FIVE

David Campbell's *A Land of Ghosts* took us by surprise in 2005, as it was based on visits he had made ten years earlier. If it took him that long to write the book, it was time well spent. He portrays nature in poetry. No more so, however, than Adrian Forsyth and Ken Miyata do in *Tropical Nature*. We tried to avoid too much reliance on their work, but it was impossible. Their book, a tour of a fascinating world that belongs to all the creatures on earth, is filled with scholarship and humor. It's a book that deserves multiple rereadings. We couldn't resist borrowing from Redmond O'Hanlon, who seems to fall on his face wherever he travels, and finds a lot to laugh about from that vantage point. Mark Plotkin is another giant of Amazon scholars, who, along with his mentor Richard Schultes, brought to an American audience a new world, rich in history and value. He has also been an enthusiastic supporter from the early days, and we're grateful for that.

48 *"All of this":* David G. Campbell, *A Land of Ghosts: The Braided Lives of People and the Forest in Far Western Amazonia* (Boston: Houghton Mifflin Company, 2005), p. 11.

49 *has described how:* Interview with Sir Ghillead Prance in *ScienceWatch*,

available at http://www.sciencewatch.com/july-aug98/science-watch_
july-aug98_page3-4.htm.

49 *"the greatest natural"*: Candice Millard, *The River of Doubt: Theodore
Roosevelt's Darkest Journey* (New York: Doubleday, 2005), p. 148.

49 *an endearing primer*: Adrian Forsyth and Kenneth Miyata, *Tropical Nature* (New York: Simon & Schuster, 1984).

49 *"one of"*: Ibid., p. 21.

49 *"the tiny dung"*: Ibid.

50 *"I'd already been"*: Scott Wallace, "Trial by Jungle," *National Geographic*, p. 127.

50 *"semiwild or wild fruits"*: Forsyth and Miyata, *Tropical*, p. 19.

51 *One methodical study*: Marcia Caldas de Castro and Burton Singer, "Migration, Urbanization and Malaria: A Comparative Analysis of Dar es
Salaam, Tanzania, and Machadinho, Rondônia, Brazil," available at
http://pum.princeton.edu/pumconference/papers/5-SingerCastro.pdf.

51 *"evolved two"*: Forsyth and Miyata, *Tropical*, p. 155.

51 *"In the Amazon"*: Edward O. Wilson, *The Diversity of Life* (Cambridge,
Mass.: Belknap Press of Harvard University Press, 1992), p. 5.

52 *"cute though they"*: Forsyth and Miyata, *Tropical*, p. 225.

52 *"overcrowded lifeboats"*: Donald Perry, *Life Above the Jungle Floor* (New
York: Simon & Schuster, 1986).

53 *scientists spotted*: available at www.estado.com.br on May 29, 2002.

53 *"duplicate the bud"*: Forsyth and Miyata, *Tropical*, p. 127.

53 *other caterpillars*: Ibid., p. 126.

53 *"reduce their own"*: Ibid., p. 132.

53 *birds' eyesight improving*: Ibid., p. 128.

53 *"manages to percolate"*: Campbell, *A Land of Ghosts*, p. 111.

53 *thirteen people*: Gareth Chetwynd, "13 Die From Rabies After Being
Bitten by Bats," *The Guardian*, April 6, 2004.

54 *"the electric eel"*: Redmond O'Hanlon, *In Trouble Again: A Journey Between the Orinoco and the Amazon* (New York: Vintage Books, 1990).

54 *a Dutch scientist*: discovered by Marc van Roosmalen, reported in *Journal do Brazil*, June 12, 2004.

54 *new species of monkey*: reported on *60 Minutes*, June 19, 2005.

55 *"We collected"*: Mark J. Plotkin, *Tales of a Shaman's Apprentice: An Eth-
nobotanist Searches for New Medicines in the Amazon Rain Forest* (New
York: Viking, 1993), pp. 225–26.

55 *created the field*: According to Plotkin, the "standard reference for eth-

nobotanists working in the Amazon" is R. E. Schultes and R. F. Raf-fauf, *The Healing Forest: Medicinal and Toxic Plants of the Northwest Amazonia* (Portland, Ore.: Dioscorides Press, 1990).

55 *During our own:* Authors' Visit, 2003.

CHAPTER SIX

Although Lovejoy, Fearnside, and Nepstad occupy a pantheon of Amazon experts, they share that place with dozens of Brazilian colleagues. By choosing the first three, we do not intend to minimize the importance of the contributions of any of the Brazilian experts. Eneas Salati and Carlos Nobre, for example, are pioneers in understanding the Amazon's role in climate change. We chose the three North Americans as a focus as much for their accessibility as for the importance of their work. They are self-effacing and would be the first to point out that they never would have been able to make the contributions they have without the help of their Brazilian colleagues. That fact is borne out in the coauthorship of most of their works, which we cite in these notes. Everton Vargas provided us with three separate interviews, and the importance we attach to his views is clear in this chapter. Finally, we acknowledge throughout this book our admiration for the humble and underappreciated Samuel Benchimol.

57 *ending August 2004: Report of the* [Brazilian] *Ministry of the Environment,* September 2005.

58 *"vicious" cycle:* In July 2006, Manaus-based scientists Antonio Nobre and Philip Fearnside issued a stark warning that the Amazon's changing climate is itself a frightening cause of deforestation. See, e.g., www.ecoearth.info.

58 *"the rain forest":* Editorial, "The Amazon at Risk," *The New York Times,* May 31, 2005.

58 *"weak, poorly co-ordinated":* Editorial, "Brazil's Landless Farmer Protest," *The Economist,* May 19, 2005.

59 *Tom Lovejoy:* See Tom E. Lovejoy et al., "Central Amazonian Forests and the Minimum Critical Size of Ecosystems Project," *Four Neotropical Rainforests* (New Haven, Conn.: Yale University Press, 1990), pp. 60–71. For further reading on Tom Lovejoy's contributions, see also:
Tom E. Lovejoy et al., "Ecological Dynamics of Tropical Forest Fragments," *Tropical Rain Forest: Ecology and Management,* Special Pub-

lication No. 2, British Ecological Society (Oxford, UK: Blackwell, 1983), pp. 377–84.

Tom E. Lovejoy et al., "Edge and Other Effects of Isolation on Amazon Forest Fragments," *The Science of Scarcity and Diversity* (Sunderland, Mass.: Sinauer, 1986), pp. 257–85.

Tom E. Lovejoy et al., "Minimum Critical Size of Ecosystems," *Forest Island Dynamics in Man-Dominated Landscapes* (New York: Springer-Verlag, 1981), pp. 7–12.

Tom E. Lovejoy et al., "Ecosystem Decay of Amazon Forest Remnants," *Extinctions* (Chicago: University of Chicago Press, 1984), pp. 295–325.

60 *"I had":* Authors' Interview, Sept. 2005.

61 *cannot be controlled:* Tom Lovejoy, "Aquertão Amazônica: Una perspectiva cientifica," *Politica Externa,* June/July/August 2005, available at www.politicaexterna.com.br.

61 *"human carrying capacity":* See Phil Fearnside, "The Effects of Cattle Pasture on Soil Fertility in the Brazilian Amazon: Consequences for Beef Production Sustainability," *Tropical Ecology* 21: 125–37 (1980). For further reading on Fearnside's contributions, see also:

Phil Fearnside, "Initial Soil Quality Conditions on the TransAmazon Highway of Brazil and Their Simulation in Models for Estimating Human Carrying Capacity," *Tropical Ecology* 25: 1–21 (1984).

Phil Fearnside, "Settlement in Rondônia and the Token Role of Science and Technology in Brazil's Amazonia Development Planning," *Interciencia* 11: 229–38 (1986).

Phil Fearnside, *Human Carrying Capacity of the Brazilian Rainforest* (New York: Columbia University Press, 1986).

Phil Fearnside, "Rethinking Continuous Cultivation in Amazonia," *Bioscience* 37: 209–14 (1987).

Phil Fearnside, "Yurimaguas Reply," *Bioscience* 38: 525–27 (1988).

Phil Fearnside, "Deforestation in Brazilian Amazon: The Rates and Causes of Forest Destruction," *The Ecologist* 19: 214–18 (1989).

Phil Fearnside, "Deforestation in Amazonia," *Environment* 31: 16–20 (1989).

Phil Fearnside, "Predominant Land Uses in Brazilian Amazonia," *Alternatives to Deforestation: Steps Toward Sustainable Use of the Amazon Rain Forest* (New York: Columbia University Press, 1990), pp. 233–51.

Phil Fearnside, "The Rate and Extent of Deforestation in Brazilian Amazonia," *Environmental Conservation* 17: 213–26 (1990).

Phil Fearnside, "Fire in the Tropical Rain Forest of the Amazon Basin," *Fire in the Tropical Biota* (Berlin, Germany: Springer-Verlag, 1990), pp. 106–16.

Phil Fearnside, "Potential Impacts of Climatic Change on Natural Forests and Forestry in Brazilian Amazonia," *Forest Ecology and Management* 78: 51–70 (1995).

Phil Fearnside, "Amazonia Deforestation and Global Warming: Carbon Stocks in Vegetation Replacing Brazil's Amazon Forest," *Forest Ecology and Management* 80: 21–34 (1996).

Phil Fearnside, "Amazonia and Global Warming: Annual Balance of Greenhouse Gas Emissions from Land-Use Change in Brazil's Amazon Region," in J. Levine (ed.), *Biomass Burning and Global Change*, Vol. 2, *Biomass Burning in South America, Southeast Asia and Temperate and Boreal Ecosystems and the Oil Fires of Kuwait* (Cambridge, Mass.: MIT Press, 1996), pp. 607–17.

Phil Fearnside, "Limiting Factors for Development of Agriculture and Ranching in Brazilian Amazonia," *Revista Brasileria de Biologia* 57: 531–49 (1997).

Phil Fearnside, "Human Carrying Capacity Estimation in Brazilian Amazonia as a Basis for Sustainable Development," *Environmental Conservation* 24: 271–82 (1997).

Phil Fearnside, "Environmental Service as a Strategy for Sustainable Development in Rural Amazonia," *Ecological Economics* 20: 53–70 (1997).

Phil Fearnside, "Greenhouse Gases from Deforestation in Brazilian Amazonia: Net Committed Emissions," *Climatic Change* 35: 321–60 (1997).

Phil Fearnside, "Protection of Mahogany: A Catalytic Species in the Destruction of Rain Forests in the American Tropics," *Environmental Conservation* 24: 303–6 (1997).

Phil Fearnside, "Monitoring Needs to Transform Amazonia Forest Maintenance into a Global Warming Mitigation Option," *Mitigation and Adaptation Strategies for Global Change* 2: 285–302 (1997).

Phil Fearnside, "Phosphorus and Human Carrying Capacity in Brazilian Amazonia," in J. P. Lynch and J. Deikman (eds.), *Phosphorus in Plant Biology: Regulatory Roles in Molecular, Cellular, Or-*

ganismic and Ecosystem Processes, American Society of Plant Physiologists (Rockville, Md., 1998), pp. 94–108.

Phil Fearnside, "Sistemas agroflorestais na política de desenvolvimento na Amazônia brasileira: Papel e limites como uso para áreas degradadas," in C. Gascon and P. Moutinho (eds.), *Floresta Amazônica: Dinâmica, Regeneração e Manejo,* Instituto Nacional de Pesquisas da Amazônia (INPA), Manaus, Amazonas, Brazil, pp. 293–312.

Phil Fearnside, "Can Pasture Intensification Discourage Deforestation in the Amazon and Pantagal Regions of Brazil?" in C. H. Wood and R. Porro (eds.), *Land Use and Deforestation in the Amazon* (Gainesville: University Press of Florida, 2002), pp. 263–364.

Phil Fearnside and R. I. Barbosa, "The Cotingo Dam as a Test of Brazil's System for Evaluating Proposed Developments in Amazonia," *Envir. Manag.* 20: 631–48 (1996).

Phil Fearnside and R. I. Barbosa, "Soil Carbon Changes from Conversion of Forest to Pasture in Brazilian Amazonia," *Forest Ecology and Management* 80: 35–46 (1996).

Phil Fearnside and W. M. Guimarães, "Carbon Uptake by Secondary Forests in Brazilian Amazonia," *Forest Ecology and Management* 108: 147–66 (1998).

Phil Fearnside et al., "Deforestation Rate in Brazilian Amazon," INPE (São Paulo: São José dos Campos, 1990).

61 *another field of study:* Authors' Interview, August 2003.

62 *Dan Nepstad:* For further reading on Nepstad's contributions, see:

Dan Nepstad et al., "The Role of Deep Roots in the Hydrolic and Carbon Cycles of Amazonian Forest and Pastures," *Nature* 372: 666–69 (1994).

Dan Nepstad et al., *Flames in the Rain Forest: Origins, Impacts and Alternatives to Amazonian Fire,* Pilot Program to Conserve the Brazilian Rain Forest (Brasília, Brazil, 1999).

Dan Nepstad et al., "A Comparative Study of Tree Establishment in Abandoned Pasture and Mature Forest of Eastern Amazonia," *Oikos* 76: 25–39 (1996).

Dan Nepstad et al., "Surmounting Barriers to Forest Regeneration in Abandoned, Highly Degraded Pastures: A Case Study from Paragominas, Pará, Brasil," *Alternatives to Deforestation: Steps Toward Sustainable Use of the Amazon Rain Forest* (New York: Columbia University Press, 1990), pp. 215–29.

Dan Nepstad et al., "Recuperation of a Degraded Amazonian Land-scape: Forest Recovery and Agricultural Restoration," *Ambio* 20: 248–55 (1991).

Dan Nepstad et al., "Large-Scale Impoverishment of Amazonian Forests by Logging and Fire," *Nature* 398: 505–8 (1999).

63 *"occupying forces"*: Authors' Interview, May 2004.

64 *"You have to"*: Authors' Interview, April 2003.

65 *"a vast lung"*: David Munk and Gareth Chetwynd, "Amazon May Be Levelled by the Humble Soya," *The Guardian*, Dec. 20, 2003.

66 *The Agriculture Ministry*: Statement issued on May 25, 2004, available at www.agriculture.gov.br.

68 *though "not one"*: Authors' Interview, Nov. 1980.

68 *"Those were years"*: Samuel Benchimol, *Amazônia: Um Pociro-Antes e Além-Depois* (Manaus: Editora Umberto Calderaro, 1977).

69 *"He felt"*: Authors' Interview, Nov. 2004.

69 *"Here are some"*: E-mail correspondence with Jaime Benchimol, December 2004.

CHAPTER SEVEN

Our notes here include multiple references to *The New York Times,* and that means Larry Rohter. We first met him in 1980 when he was a correspondent for *Newsweek* in Rio. We can't think of a *Times* reporter who has been in a foreign bureau as long, and we assume that's because his newspaper recognizes the same thoroughness, clarity, and quality that we find in his work. Luis Bitencourt, who contributed to *Environment and Security in the Amazon Basin,* capably headed for many years the Brazil Center at the Woodrow Wilson International Center for Scholars. The center is a resource for information about current trends in Brazil and has a regular menu of informative speakers whose presentations have provided helpful background. Most of the history of the cocaine trade comes from our interviews with Mauro Sposito and files he provided us. Jonathan Kandell has an excellent account of narco-trafficking in *Passage Through El Dorado.*

71 *Built at a cost:* See Thomaz Guedes da Costa, "SIVAM: Challenges to the Effectiveness of Brazil's Monitoring Project for the Amazon," and Clóvis Brigagão, "SIVAM: Environmental and Security Monitoring in Amazônia," found in *Environment and Security in the Amazon Basin,* pp. 99 and 115, respectively.

72   *When SIVAM became:* Larry Rohter, "Brazil Employs Tools of Spying to Guard Itself," *The New York Times,* July 27, 2002.

72   *Young scientists working:* Authors' Visit, August 2003.

72   *Braga of Amazonas told us:* Authors' Interview, August 2003.

73   *In November 2003:* reprinted at www.greenpeace.org.br, Nov. 20, 2003.

74   *This reservation had:* "Diamonds of Discord," *Brazzil,* April 2002; available at http://www.brazzil.com/content/view/6436/38/.

74   *the Associated Press reported:* available at www.indianz.com/news/2004.

74   *Reuters reported:* available at www.plantetark.com/avantgo/dailynewsstory.

75   *Mauro Sposito:* Authors' Visit, Nov. 2003.

76   *The king had:* For detailed description of narco-trafficking, see Jonathan Kandell, *Passage Through El Dorado* (New York: W. Morrow, 1984), pp. 231–44.

79   *Sposito sees himself:* Larry Rohter, "Latest Battleground in Latin Drug War: Brazilian Amazon," *The New York Times,* Oct. 30, 2000.

CHAPTER EIGHT

Ademir Braz, a local poet, served as our guide on our visits with Major Curio. Braz's latest collection of poetry is called *Rebanho de Patras & Esta Terra,* and we wish him success with it. David Cleary's book on gold mining is a great source. Cleary himself, a resident of Brasília, is a student of all things Brazilian, as can be seen in his authorship of the *Lonely Planet* guidebook to the country.

83   *"as rainfall changes":* David Cleary, *Anatomy of the Amazon Gold Rush* (Iowa City: University of Iowa Press, 1990).

84   *"The exact circumstances":* Ibid., p. 169.

84   *José Maria da Silva:* Authors' Interview, Nov. 1980.

85   *Antonio Gomes Souza:* Authors' Interview, Nov. 1980.

85   *Joaquin Almeida:* Authors' Interview, Nov. 1980.

86   *"economic dislocation":* Cleary, *Anatomy of the Amazon,* p. 171.

87   *Our own story:* *Parade* magazine, March 30, 1981.

89   *Many miners:* Authors' Visit, Nov. 1980.

89   *Born in the interior:* Larry Rohter, "A Man of Many Names but One Legacy in the Amazon," *The New York Times,* Sept. 11, 2004.

89   *Several thousand troops:* Larry Rohter, "Long After Guerilla War, Sur-

vivors Demand Justice from Brazil's Government," *The New York Times,* March 28, 2004.

90   *Looking back:* Authors' Interview, May 2004.

92   *According to Amnesty International:* The scope of the incident and the uncertainty as to number of victims is reported at "Appendix A: The Tocantin Bridge Massacre, Pará, 29 December 1987," available at www.web.amnesty.org.

92   *When we visited:* Larry Rohter, "Brazilian Miners Wait for Payday After Diet of Bitterness," *The New York Times,* Aug. 23, 2004.

CHAPTER NINE

96   *"The majority of":* Authors' Interview, Nov. 1980.

96   *"We have the technology":* Authors' Interview, Dec. 1981.

97   *"As a general rule":* Authors' Interview with Jim Cleveland, Dec. 1981.

98   *It's estimated that there are:* E-mail confirmation from Sven Wolff, general manager, Petrobras facility at Urucu, March 2, 2006.

98   *"This is not Saudi Arabia":* Authors' Interview, Nov. 2003.

98   *All gasoline sold:* Dan Morgan, "Brazil's Biofuel Strategy Pays Off as Gas Prices Soar," *The Washington Post,* June 18, 2005.

99   *The hydrocarbons:* Cecilia Smith, "Taking Root in the Jungle," *Houston Chronicle,* June 22, 2005.

99   *"You could practically":* Authors' Interview, Nov. 2003.

99   *The daily output:* Smith, "Taking Root."

100   *"If people want":* "Amazon Gas Heralds Change in Brazilian Rain Forest," *Reuters,* December 20, 2004.

100   *"We want to make":* Authors' Interview, Nov. 2003.

100   *The eighteen hundred workers:* Authors' Visit, Nov. 2003.

CHAPTER TEN

Our account of the Jari Project is based in part on the almost two weeks we spent there in 1980 and 1981 at a time when the place was closed to most journalists. We've been to Porto Velho several times to visit the remnants of the Madeira-Mamoré Railroad. We spent several days in Ford's Belterra, although we never made it all the way to Fordlandia. We've been to the Opera House in Manaus as often as to our neighborhood movie theater.

102   *Charles Wagley:* Charles Wagley, *Amazon Town: A Study of Man in the Tropics* (New York: Oxford University Press, 1976), p. 65.

105   *Few of Sinop's:* Authors' Visit, Nov. 1981 and Nov. 2003.

105   *In Porto Velho:* Authors' Visit, Nov. 1981 and Nov. 2003.

105   *"Before rubber":* Jonathan Kandell, *Passage Through El Dorado* (New York: W. Morrow, 1984), p. 91.

106   *They built:* Larry Rohter, "Adventures in Opera: A 'Ring' in the Rain Forest," *The New York Times,* May 9, 2005.

107   *a self-made billionaire:* An excellent overview of the project as it began to erode can be found at: Gwen Kinkead, "Trouble in D.K. Ludwig's Jungle," *Fortune,* April 20, 1981.

108   *The Jari Project:* Larry Rohter, "A Mirage of Amazonian Size; Delusions of Economic Grandeur Deep in Brazil's Interior," *The New York Times,* Nov. 9, 1999.

110   *"I arrived here":* Authors' Interview, Nov. 2003.

111   *"the promise of a better life":* Authors' Interview, Nov. 2003.

CHAPTER ELEVEN

Marina Silva is so well known in the international environmental movement that there's no shortage of material available on the Internet or through NGO or Brazilian government sources. The difficulty is in trying to find consistency in some of these accounts. We found the biography of Marina Silva by Ziporah Hildebrandt to be helpful. We had three separate meetings with Marina Silva, one in Brasília and two in Washington—a brief discussion and then a more lengthy one. We recognize the role played by Mary Allegretti and Steve Schwartzman in the Chico Mendes era. Ms. Allegretti served as secretary of the Amazon under Marina Silva and now is resident at the University of Florida. We saw her in Washington during a visit with Governor Jorge Viana, and she provided helpful feedback. Dr. Schwartzman is another member of the pantheon of Amazon experts, having been involved in a variety of issues over the last thirty years. He was most helpful in pointing us in the direction of *Terra do Meio* as the site of the most serious violence and providing us with a list of contacts. He was instrumental in the process resulting in Tascísio Feitosa's winning the 2006 Goldman Environmental Prize, an enormously important achievement not only for the recognition of his work but for the security the renown will provide Feitosa. Foster

Brown's *Innocence in Brazil* is a diary of a modern-day Indiana Jones. It's available to his pen pals; perhaps someday a publisher will come along and help him share it with the wider audience it deserves.

115 *"the highest-class":* Authors' Interview, Nov. 1980 and Nov. 1981. Description from Brian Kelly and Mark London, *Amazon* (New York: Harcourt Brace Jovanovich, 1983), p. 45.

116 *Born in 1958:* Ziporah Hildebrandt, *Marina Silva: Defending Rain Forest Communities in Brazil* (New York: Feminist Press at the City University of New York, 2001). Various handouts from Brazil's Ministry of the Environment also provided background information.

117 *"fifteen hundred":* Alex Shoumatoff, *The World Is Burning: Murder in the Rain Forest* (Boston: Little, Brown and Company, 1990), p. 68.

117 *"The cycle of":* Andrew Revkin, *The Burning Season: The Murder of Chico Mendes and the Fight for the Amazon Rain Forest* (Boston: Houghton Mifflin Company, 1990), p. 162.

119 *She spoke to us:* Authors' Visit, April 2003.

120 *"This region can't":* Matt Moffett, "Fading Green: Brazil's President Sees New Growth in the Rain Forest," *The Wall Street Journal,* Oct. 16, 2003.

123 *Governor Edmundo Pinto:* Larry Rohter, "A Brazilian Campaign That Is All About the Jungle," *The New York Times,* Sept. 23, 2002.

124 *"His progressive views":* Laurie Goering, "In an Endangered Amazon, Seeds of Hope Taking Root," *Chicago Tribune,* June 28, 1999.

124 *Viana pointed out:* Authors' Follow-up Interview, Dec. 2005.

124 *the actual number:* confirmed in Larry Rohter, "Discovering Amazon's Rain Forest's Silver Lining," *The New York Times,* Sept. 10, 2002.

124 *Viana's fund-raising success:* Authors' Interview, Aug. 2003.

125 *"In 2002":* Authors' Interview, Sept. 2005.

126 *The governor of the neighboring state:* Authors' Interview with Valerio Gomes, Sept. 2005.

128 *The doctoral student:* Authors' Interview, Sept. 2005.

128 *Irving Foster:* Authors' Visits, Aug. 2003 and Sept. 2005.

129 *environmentally delicate:* "Estrada do Rio Branco, Acre, Brasil dos Portos do Pacifico: Como maximizar os benefícios e minimizar os prejuízos para o desenvolvimento sustentável da Amazônia Sul-Ocidental," Brown, Brilharte, Mendoza e Ribeiro de Oliveira, *Encuentro Internacional de Integracion Regional—Bolivia, Brasil y Peru* (Editora CEPEI, Lima, 2002).

131   *The fabled Xapuri:* Authors' Visit, Aug. 2003.

133   *Jerry Correia:* Authors' Visit, Aug. 2003.

CHAPTER TWELVE

We interviewed Scott Paul of Greenpeace in Washington, D.C., who provided a helpful summary of Greenpeace's efforts in curtailing the mahogany trade. We sought to interview Paulo Adario, the head of Greenpeace in Manaus, but his security precautions and our lack of advance work prevented that. Greenpeace's work in this area has been extremely effective. A series of reports from the fall of 2001 document this work. We found the well-resonated *Outside* piece by Patrick Symmes to be an invaluable resource.

135   *In the town of:* Authors' Visit, May 2004.

135   *Greenpeace has made:* "Partners in Mahogany Crime," Greenpeace, Sept.–Oct. 2001.

138   *how the tree increases:* Patrick Symmes, "Blood Wood," *Outside* magazine, Oct. 2002.

139   *"Mahogany starts":* Authors' Interview, Sept. 2003.

140   *"Fueled by high":* "Partners in Mahogany Crime," Greenpeace.

142   *undercover federal marshal:* Authors' Interview, Nov. 1980.

143   *children of immigrants:* Authors' Interview, May 2004.

CHAPTER THIRTEEN

The 2003 World Bank report by Sergio Margulis, *Causes of Deforestation of the Brazilian Amazon,* changed the popular perception that cattle ranching was an unprofitable business. In many ways, it's as significant in this area as Roosevelt's work was in anthropology. We also refer to the informative work on slavery by Binka Le Breton, who seems to hang out at home in southern Brazil until boredom strikes, then sets out for adventure by herself or with her daughter. In addition to *Trapped: Modern-day Slavery in the Brazilian Amazon,* she has written about the Amazon in *Voices from the Amazon, A Land to Die For,* and *Rainforest* (London: Longmans, 1997).

147   *"We gazed out":* Brian Kelly and Mark London, *Amazon* (New York: Harcourt Brace Jovanovich, 1983), p. 231.

147   *"Livestock became":* Susanna Hecht and Alexander Cockburn, *The Fate*

of the Forest: Developers, Destroyers, and Defenders of the Amazon (London: Verso, 1989), p. 149.

147 *"McDonald's has been"*: See www.mcspotlight.org and www.forests .org, "McDonald's Linked to Rainforest Destruction."

148 *None of this*: "Hamburger Connection Fuels Amazon Destruction," David Kaimowitz, Benoit Mertens, Sven Wunder, and Pablo Pacheco, CIFOR (Center for International Forestry Research).

148 *"Several studies"*: Hecht and Cockburn, *The Fate*, p. 150.

148 *from 1990 to 2005*: Maria del Carmen Vera Diaz and Stephan Schwartzman, "Carbon Offsets and Land Use in the Brazilian Amazon," in Paulo Moutinho and Stephan Schwartzman (eds.), *Tropical Deforestation and Climate Change* (Belém: Amazon Institute for Environmental Research, 2005).

148 *amount of deforested land*: Sergio Margulis, "Causes of Deforestation of the Brazilian Amazon" (World Bank, 2003), p. 19; INPE statistics for 2005.

148 *With a 5.5 to 1 ratio*: "Hamburger Connection." According to the most recent census figures available, the area of land devoted to crops in 1995–96 amounted to 5,608,000 hectares, while the figure for pasture was 33,579,000.

148 *this change in economics*: Margulis, "Causes of Deforestation," p. 11.

149 *We traveled to the*: Authors' Visits, May 2004.

150 *Between 1997 and 2003*: "Hamburger Connection," citing Margulis and USDA, "Brazil Livestock and Products: Semiannual Report," 2004. GAIN report BR4605.

151 *road networks doubled*: Margulis, "Causes of Deforestation," p. 6. See also K. Chomitz and T. S. Thomas, "Geographic Patterns of Land Use and Land Intensity Development Research Group" (Washington, D.C., 2000); D. Nepstad et al., *Roads in the Rainforest: Environmental Costs for the Amazon* (Belém: MGM Gráfica e Editora, 2002); G. O. Carvalho et al., "Frontier Expansion in the Amazon: Balancing Development and Sustainability," *Environment* 44(3): 34–46.

152 *3,937 children were shot*: available at www.amnestyusa.org/lordofwar/ finalchildren.html.

152 *"Deforestation is taking place undetected"*: Margulis, "Causes of Deforestation," p. 24.

153 *In 2003*: Binka Le Breton, *Trapped: Modern-day Slavery in the Brazilian Amazon* (Bloomfield, Conn.: Kumarian Press, Inc., 2003).

153   *The Brazilian writer:* Jonathan Kandell, *Passages Through El Dorado* (New York: W. Morrow, 1984), p. 95.

154   *twenty-five thousand slaves:* Larry Rohter, "Brazilian Leader Introduces Program to End Slave Labor," *The New York Times,* March 14, 2003.

154   *"The experience of":* Margulis, "Causes of Deforestation," p. 53.

155   *moved cattle ranching:* See, for example, Stephen A. Vosti, Chantal Line Carpentier, Julie Witcover, and Judson F. Valentim, "Intensified Small-scale Livestock Systems on the Western Brazilian Amazon," *Agricultural Technologies and Tropical Deforestation,* CAB International 2001, p. 113.

156   *"We can raise":* Authors' Interview, August 2003.

157   *"Cattle ranching":* Authors' Interview, Sept. 2005.

158   *a five-thousand-hectare:* Authors' Visit, Aug. 2003.

CHAPTER FOURTEEN

Governor Maggi and his staff made themselves accessible throughout this project, and we were able to follow up interviews with e-mail inquiries. We relied on the biography of the Maggi family, *Corporacao e Rede em Areas de Fronteira.* The annual USDA Foreign Agricultural Service reports were reliable sources for the growth of the Brazilian soybean industry. George Flaskerud of North Dakota State University in July 2003 produced the report about the component costs of soybean production, showing the competitive edge shifting to the Brazilian Amazon: "Brazil's Soybean Production and Impact." We received the information about the capacities of the Maggi facility in Porto Velho and at Itacoatiara during our on-site visits.

162   *where Maggi agreed:* Authors' Visit, Aug. 2003.

162   *"in the Brazilian":* Mac Margolis, *The Last New World: The Conquest of the Amazon Frontier* (New York: W. W. Norton and Company, 1992), p. 65.

163   *We showed him:* Brian Kelly and Mark London, *Amazônia: Um Grito de Alerta* (Rio de Janeiro: Editora Record, 1983).

165   *The international media:* Michael McCarthy and Andrew Buncombe, "The Rape of the Rainforest . . . and the Man Behind It," *The Independent,* May 20, 2005; available at http://news.independent.co.uk/environment/article222264.ece.

165   *In the months after:* Authors' Visit, Nov. and Dec. 2003.

167 *Maggi's father pursued:* Carlos Alberto Franco da Silva, *Grupo André Maggi: Corporação e Rede em Áreas de Fronteira* (Cuiabá: Entrelinhas, 2003).

168 *promoted cooperation:* background information available in Larry Rohter, "Relentless Foe of the Amazon Jungle: Soybeans," *The New York Times,* Sept. 17, 2003.

168 *"It is apparent":* USDA Foreign Agricultural Service, "Brazil: Future Agricultural Expansion Potential Underrated," Jan. 21, 2003, p. 1; available at http://www.fas.usda.gov/pecad2/highlights/2003/01/Ag_expansion/.

168 *The president of:* Renee Anderson, "Brazil's Crop Expansion, Infrastructure Improvements Send Shock Waves North"; available at http://server.admin.state.mn.us/resource.html?Id=1125.

168 *"I went down":* Jerry Carlson, "Iowa Soybeans Grower Tim Burrock: Against Brazil, We're not Even on the Global Game," *ProFarmer* magazine, Feb. 6, 2001.

169 *annual global soy:* George Flaskerud, "Brazil's Soybean Production and Impact," North Dakota State University Extension Service, July 2003, p. 14; available at http://www.ext.nodak.edu/extpubs/plantsci/rowcrops/eb79w.htm.

169 *World production in 2005:* USDA, Foreign Agricultural Service, "World Agricultural Production," June 2005.

169 *By most estimates:* Lester R. Brown, *Outgrowing the Earth: The Food Security Challenge in the Age of Falling Water Tables and Rising Temperatures* (New York: W. W. Norton and Company, 2005), p. 161.

169 *Brazil earned ten billion dollars:* "The American Soybean Association Weekly Update," April 2005; available at www.grains.org/galleries/asa_weekly/soy041105.pdf.

169 *China's rapidly growing:* The Dow Jones-AIG Commodity Index Daily Commentary, Feb. 4, 2004; available at http://www.djindexes.com/mdsidx/views/Products/Aig/html/aig/AIG_archive/DJAIG_Feb04_04.pdf.

169 *China's imports:* USDA Foreign Agricultural Service, Global Agricultural Information Network ("GAIN"), GAIN Report # BR3020, Oct. 28, 2003, p. 26; available at http://www.fas.usda.gov/gainfiles/200310/145986620.pdf.

169 *63 percent of European:* Van Maarten Dros, "Managing the Soy Boom: Two Scenarios of Soy Production Expansion in South America," commissioned by WWF, June 2004, p. 8.

169 *"The combination of"*: USDA Production Estimates and Crop Assessment Division, Foreign Agricultural Service, "The Amazon: Brazil's Final Soybean Frontier," Jan. 13, 2004, p. 7; available at http://www.fas.usda.gov/pecad/highlights/2004/01/Amazon/Amazon_soybeans.htm.

169 *"Soybean production in 2003"*: Flaskerud, "Brazil's Soybean Production and Impact," p. 12.

170 *total cost in Rotterdam:* Flaskerud, "Brazil's Soybean Production and Impact," p. 12.

170 *2003 soybean variety:* Anne Casson, "Oil Palm, Soybeans & Critical Habitat Loss," WWF Forest Conversion Initiative, Aug. 2003, p. 12 footnote 12; available at http://assets.panda.org/downloads/oilpalm-soybeanscriticalhabitatloss25augusto3.pdf.

170 *from 1997 to 2003:* USDA Production Estimates and Crop Assessment Division, Foreign Agricultural Service, "The Amazon: Brazil's Final Soybean Frontier," Jan. 13, 2004, p. 7; available at http://www.fas.usda.gov/pecad/highlights/2004/01/Amazon/Amazon_soybeans.htm.

170 *"Right now"*: Editorial, "The Amazon at Risk," *The New York Times,* May 31, 2005.

171 *productivity of soy:* Ulrike Bickel and Van Maarten Dros, "The Impacts of Soybean Cultivation on Brazilian Ecosystems," p. 14, commissioned by WWF, Oct. 2003.

171 *"world's wealthiest"*: Elizabeth Becker, "Western Farmers Fear Third-World Challenge to Subsidies," *The New York Times,* Sept. 9, 2003.

172 *Thirty-two-year-old:* Authors' Interview, May 2005.

173 *frustrated environmentalists:* Ibid.

174 *In a separate interview:* Authors' Interview, May 2004.

175 *In June 2005:* "Dozens Nabbed in Brazil Logging Crackdown," Associated Press, June 3, 2005.

175 *Department of Agriculture:* "The Amazon: Brazil's Final Soybean Frontier," Production Estimates and Crop Assessment Division, Foreign Agricultural Service, Jan. 13, 2004, p. 5.

177 *His vision:* Authors' Visit, Nov. 2003.

178 *Maggi quietly:* Authors' Visit, Aug. 2003.

179 *IPAM has predicted:* Dan Nepstad and J.J.P. Capobianco, "Roads in the Rainforest: Environmental Costs for the Amazon" (Belém: Instituto de Pesquisa de Amazonia and Instituto Socioambiental, 2002), and World Bank, Global Economic Prospects and the Developing

Countries 2001; available at http://www.worldbank.org/poverty/data/trends/inequal.htm.

180 *While the environmentalists:* "The Amazon Will Be Occupied," interview of David McGrath in *Veja* magazine by Leonardo Contino, Nov. 12, 2003.

180 *"In the last":* Authors' Interview, May 2004.

181 *"We have enough":* Maggi speech in Washington, D.C., Dec. 2003.

182 *"The millions of":* Maggi titles his speech "A Agricultura e o Império do Medo" ("Agriculture and the Empire of Fear"), Oct. 2001.

183 *In September 2004:* Elizabeth Becker and Todd Benson, "Brazil's Road to Victory Over U.S. Cotton," *The New York Times*, May 4, 2005.

CHAPTER FIFTEEN

The interviews in this chapter come from a mix of visits by the authors and by the researchers Amy Rosenthal and Vera Reis.

186 *"The summit also":* Mac Margolis, *The Last New World: The Conquest of the Amazon Frontier* (New York: W. W. Norton and Company, 1992), p. 95.

186 *As it leaves:* Authors' Visits, Dec. 2003 and May 2005.

188 *C. R. Almeida:* Larry Rohter, "Brazil's Lofty Promises After Nun's Killing Prove Hollow," *The New York Times*, Sept. 23, 2005.

189 *Cordiality is:* Joseph A. Page, *The Brazilians* (New York: Da Capo Press, 1995), p. 9.

190 *Run by Luzia:* Authors' Visits, Nov. 2003 and May 2005.

191 *Tarcisio Feitosa:* Authors' Visits, Nov. 2003 and May 2005. In April 2006, Feitosa won the well-deserved Goldman Environmental Prize, which awards international renown and $125,000.

194 *shrugged off his handicaps:* Authors' Interview, Nov. 2003.

198 *"the innate differences":* Jared Diamond, *Guns, Germs, and Steel: The Fates of Human Societies* (New York: W. W. Norton and Company, 1999), p. 405.

198 *Why was Pula Pula:* Jeffrey Kluger, "Ambition: Why Some People Are Most Likely to Succeed," *Time*, Nov. 14, 2005.

199 *"Some writers":* David McGrath, "Regatão and Caboclo: Itinerant Traders and Smallholder Resistance in the Brazilian Amazon," in Stephen Nugent and Mark Harris (eds.), *Some Other Amazonians:*

Perspective on Modern Amazonia (London: Senate House, 2004), p. 184.

199 *"Janus-face":* McGrath, "Regatão and Caboclo," p. 178.

200 *global consumerism:* Authors' Visit, Dec. 2003.

204 *"[One] could not":* Candice Millard, *The River of Doubt: Theodore Roosevelt's Darkest Journey* (New York: Doubleday, 2005), p. 194.

CHAPTER SIXTEEN

Again, the interviews in this chapter come from a mix of visits by the authors and by researchers Amy Rosenthal and Vera Reis.

207 *"a poor man's road":* Marianne Schmink and Charles H. Wood, *Contested Frontiers in Amazonia* (New York: Columbia University Press, 1992), p. 4.

208 *"By the 1970's":* Jonathan Kandell, *Passage Through El Dorado* (New York: W. Morrow, 1984), p. 109.

208 *"no resident":* Hernando de Soto, *The Other Path: The Economic Answer to Terrorism* (New York: Basic Books, 1989), p. 180.

208 *"many disputes":* "Ronald Coase and the Coase Theorem," p. 2; available at http://www.aliveness.com/kangaroo/L-chicoase.htm.

208 *Daniel Webster:* Daniel Webster, "First Settlement of New England," speech delivered at Plymouth, Massachusetts, Dec. 22, 1820, to commemorate the two-hundredth anniversary of the landing of the Pilgrims at Plymouth. *The Writings and Speeches of Daniel Webster,* vol. 1, p. 214 (1903).

209 *with the law:* Alexis de Tocqueville (ed. Richard D. Heffner), *Democracy in America* (New York: New American Library, 1956), p. 107.

210 *"where the law":* quoted from *O Estado de S. Paulo* (Sept. 2003) in "State of Conflict: An Investigation into the Landgrabbers, Loggers and Lawless Frontiers in Pará State, Amazon," published by Greenpeace and available at www.greenpeace.org.

210 *"For migrant families":* Schmink and Wood, *Contested Frontiers,* p. 276.

210 *"It is no wonder":* Ibid., p. 342.

211 *Clement owns:* Authors' Visit, Dec. 2003.

212 *2,738,085 hectares:* Schmink and Wood, *Contested Frontiers,* p. 262.

212 *3,262,960 hectares:* Ibid., p. 269.

213 *José Gomes:* Authors' Visit, May 2004.

214 *His small office:* Authors' Visit, May 2004.

215 *The CPT representative:* Authors' Visit, Dec. 2000.

215 *Church listened:* Leonardo Boff and Cledoris Boff, *Introducing Liberation Theology* (London: Burns & Dates, 1987).

215 *"The Church is":* Anthony Grafton, "Reading Ratzinger," *The New Yorker,* July 25, 2005.

216 *two French priests:* They were Aristeder Camio and Francisco Gourion. We followed their story on our early visits to the Amazon. Also noted at Schmink and Wood, *Contested Frontiers,* p. 183.

217 *"lived among those":* Andrew Buncombe, "The Life and Brutal Death of Sister Dorothy, a Rainforest Martyr," *The Independent,* Feb. 15, 2005.

217 *We tried to:* Authors' Visit, May 2005.

218 *According to her brother:* Interview with David Stang, available at www.maryknoll.org.

218 *Ulisses Tavares:* Authors' Visit, May 2005.

219 *According to the eyewitness:* Tavares interview, ibid.

220 *"The story is":* Larry Rohter, "Brazil Promises Crackdown After Nun's Shooting Death," *The New York Times,* Feb. 14, 2005.

220 *During the weeks:* Larry Rohter, "Brazil's Lofty Promises After Nun's Killing Prove Hollow," *The New York Times,* Sept. 23, 2005.

220 *Nationalists decried:* Janer Christaldo, "All Brazil Needed: An American Martyr!" *Brazzil* magazine, Feb. 24, 2005.

220 *He personally:* Grafton, "Reading Ratzinger."

221 *"We have lost":* Andrew Buncombe, "Brazil: Battle for the Heart of the Rainforest," *The Independent,* March 29, 2005.

221 *Itaituba:* Authors' Visit, Dec. 2000 and Dec. 2003.

222 *Maria Elza Ezequiel:* Authors' Interview, Dec. 2003.

223 *Many of the:* Authors' Interview, Dec. 2003.

225 *Charles Trocate:* Authors' Interview, May 2004.

227 *According to Raul Jungman:* Authors' Interview, Aug. 2003.

229 *Arlindo Alves:* Authors' Interview, May 2004.

230 *Four structures:* Authors' Visit, May 2004.

231 *They received assurances:* Authors' Visit, May 2004.

232 *who promptly condemned it:* Fernando Henrique Cardoso, *The Accidental President of Brazil: A Memoir* (New York: PublicAffairs, 2006), p. 208.

232 *Amnesty International:* Its response to the incident, investigation, and prosecution can be seen at www.web.amnesty.org.

234 *"I knew that":* Cardoso, *The Accidental President of Brazil,* p. 208.

234 *The March 26th Settlement:* Authors' Visit, May 2004.

CHAPTER SEVENTEEN

We interviewed Governor Braga three times, and have combined the substance of those interviews in this chapter.

241 *In the meantime:* Authors' Interview, Aug. 2003.

243 *More than one hundred thousand:* SUFRAMA presentation at the Harvard Club in New York City, March 23, 2006; publications available at http://www.suframa.gov.br.

245 *to succeed he knew:* Authors' Interview, Aug. 2003.

249 *Recently:* Paulo Pinda, "Poisonous Tree Frog Could Bring Wealth to Tribe in Brazilian Amazon," *The New York Times,* May 30, 2006.

249 *Three businesses:* Authors' Visits, Aug. 2003, Nov. 2003, and May 2004.

250 *"plays a key role":* Gregory Prang, "Social and Economic Change in Amazonia: The Case of Ornamental Fish Collection in the Rio Negro Basin," in *Some Other Amazonians: Perspective on Modern Amazonia* (London: Senate House), p. 71.

251 *Amaral's family:* Larry Rohter, "A Quest to Save a Tree, and Make the World Smell Sweet," *The New York Times,* Aug. 30, 2005.

254 *"It's very difficult":* Authors' Visit, Oct. 2003.

254 *"People don't pollute":* Interview of Ronald Coase in *Reason* magazine, Jan. 1997, p. 4; available at http://www.reason.com/9701/int.coase .shtml.

EPILOGUE

The journey on BR-319 took place in August 2004.

Additional Sources

In addition to the sources appearing in our notes, we recommend the following books, which we relied upon for general background information:

Alston, Lee J., Gary D. Libecap, and Bernardo Mueller, *Titles, Conflict, and Land Use: The Development of Property Rights and Land Reform on the Brazilian Amazon Frontier* (Ann Arbor: The University of Michigan Press, 1999).

Benchimol, Samuel, *Amazônia: Um pouco-antes e além-depois* (Manaus: Editora Umberto Calderaro, 1977).

Bierregaard, Robert O., Jr., Claude Gascon, Thomas E. Lovejoy, and Rita Mesquita (eds.), *Lessons from Amazonia: The Ecology and Conservation of a Fragmented Forest* (New Haven, Conn.: Yale University Press, 2001).

Browder, John O., and Brain J. Godfrey, *Rainforest Cities: Urbanization, Development, and Globalization of the Brazilian Amazon* (New York: Columbia University Press, 1997).

Caulfield, Catherine, *In the Rainforest* (Chicago: University of Chicago Press, 1984).

Cohen, J. M., *Journeys Down the Amazon: Being the Extraordinary Adventures and Achievements of the Early Explorers* (London: Charles Knight & Company, 1975).

Cowell, Adrian, *The Decade of Destruction: The Crusade to Save the Amazon Rain Forest* (New York: Henry Holt, 1990).

De Onis, Juan, *The Green Cathedral: Sustainable Development of Amazonia* (New York: Oxford University Press, 1992).

Dos Passos, John, *Brazil on the Move* (Garden City, N.Y.: Doubleday & Co., 1963).

Hall, Anthony (ed.), *Amazonia at the Crossroads: The Challenge of Sustainable Development* (London: Institute of Latin American Studies, 2000).

Hardin, Garrett, "The Tragedy of the Commons," *Science* 162: 1243–48 (1968) (journal article).

Hemming, John, *Amazon Frontier: The Defeat of the Brazilian Indians* (Cambridge, Mass.: Harvard University Press, 1987).

Hemming, John, *Red Gold: The Conquest of the Brazilian Indians* (Cambridge, Mass.: Harvard University Press, 1978).

Kane, Joe, *Running the Amazon* (New York: Alfred A. Knopf, 1989).

Laurance, W., and R. Bierregaard, Jr. (eds.), *Tropical Forest Remnants: Ecology, Management, and Conservation of Fragmented Communities* (Chicago: University of Chicago Press, 1997).

Little, Paul E., *Amazonia: Territorial Struggles on Perennial Frontiers* (Baltimore, Md.: Johns Hopkins University Press, 2001).

Mahar, Dennis, *Frontier Development Policy in Brazil: A Study of Amazonia* (New York: Praeger, 1979).

Mahar, Dennis, *Government Policies and Deforestation in Brazil's Amazon Region* (Washington, D.C.: The World Bank, 1989).

Matthiessen, Peter, *At Play in the Fields of the Lord* (New York: Vintage Books, 1987).

Mendes, Chico, *Fight for the Forest: Chico Mendes in His Own Words* (London: Latin American Bureau, 1989).

Moran, Emilio, *Developing the Amazon* (Bloomington: Indiana University Press, 1981).

Myers, Norman, *The Primary Source: Tropical Forests and Our Future* (New York: W. W. Norton, 1984).

Raffles, Hugh, *In Amazonia: A Natural History* (Princeton, N.J.: Princeton University Press, 2002).

Seuss, Dr., *The Lorax* (New York: Random House for Young Readers, 1971).

Stone, Roger D., *Dreams of Amazonia* (New York: Viking, 1985).

Stone, Roger D., *The Nature of Development: A Report from the Rural Tropics on the Quest for Sustainable Economic Growth* (New York: Alfred A. Knopf, 1992).

Weart, Spencer R., *The Discovery of Global Warming* (Cambridge, Mass.: Harvard University Press, 2003).

INDEX

—

ABOUT THE AUTHORS

MARK LONDON is a trial lawyer in Washington, D.C. BRIAN KELLY is the executive editor of *U.S. News & World Report*. They are the authors of *Amazon* and *The Four Little Dragons*.

ABOUT THE TYPE

This book was set in Scala, a typeface designed by Martin Majoor in 1991. It was originally designed for a music company in the Netherlands and then was published by the international type house FSI FontShop. Its distinctive extended serifs add to the articulation of the letterforms to make it a very readable typeface.